COVID-19 and International Development

Elissaios Papyrakis
Editor

COVID-19 and International Development

Editor
Elissaios Papyrakis
International Institute of Social Studies (ISS)
Erasmus University Rotterdam
The Hague, The Netherlands

ISBN 978-3-030-82341-2   ISBN 978-3-030-82339-9 (eBook)
https://doi.org/10.1007/978-3-030-82339-9

© The Editor(s) (if applicable) and The Author(s), under exclusive license to Springer Nature Switzerland AG 2022
This work is subject to copyright. All rights are solely and exclusively licensed by the Publisher, whether the whole or part of the material is concerned, specifically the rights of translation, reprinting, reuse of illustrations, recitation, broadcasting, reproduction on microfilms or in any other physical way, and transmission or information storage and retrieval, electronic adaptation, computer software, or by similar or dissimilar methodology now known or hereafter developed.
The use of general descriptive names, registered names, trademarks, service marks, etc. in this publication does not imply, even in the absence of a specific statement, that such names are exempt from the relevant protective laws and regulations and therefore free for general use.
The publisher, the authors and the editors are safe to assume that the advice and information in this book are believed to be true and accurate at the date of publication. Neither the publisher nor the authors or the editors give a warranty, expressed or implied, with respect to the material contained herein or for any errors or omissions that may have been made. The publisher remains neutral with regard to jurisdictional claims in published maps and institutional affiliations.

Cover image: 'Horns of plenty' (oil on canvas, 2020) © Peeter Burgeik

This Springer imprint is published by the registered company Springer Nature Switzerland AG
The registered company address is: Gewerbestrasse 11, 6330 Cham, Switzerland

Preface

The Covid-19 pandemic is currently reversing development gains that were achieved with much dedicated effort in most parts of the developing world during the last few decades. The Covid-19 economic downturn contributed to a rise in global poverty for the first time since the early 2000s. Existing inequalities interlink with the Covid-19 impacts to intensify the vulnerabilities of low-income communities and individuals. Many intertwined factors generate a toxic mix that disadvantages developing economies and their populations. The unprecedented health crisis in India (especially since April 2021) with severe oxygen shortage at hospitals exposed the unpreparedness of its public health system. While developed economies can resort to significant fiscal and monetary expansion (at a very low cost), developing nations face severe liquidity constraints and restricted access to international funds. At the same, governments in developed nations reduced their bilateral development assistance reserving funds for their own economies. This lack of international solidarity is further exemplified by the unequal distribution of vaccines globally.

The media's portrayal of the Covid-19 pandemic in developed countries is largely skewed towards highlighting domestic health developments and policy responses. In an ever-interconnected world, however, isolationist reactions are unlikely to provide solutions to global health problems that will prevent the further mutation of the virus and its rapid spread across national borders. Similarly, protectionist measures and curbed international aid are likely to slow down the pace of the global economic recovery. This is a book written by development specialists to explicitly discuss how the current pandemic influences international development debates and to reflect on the specificities of developing countries and the challenges they currently face. Its novelty lies in being the first book that explains the multifaceted implications of the ongoing Covid-19 pandemic for international development from a social science perspective. It is the outcome of a collaborative project between scholars based at the International Institute of Social Studies, Erasmus University Rotterdam, and other scientists coming primarily from the Global South. It is a joint effort to highlight inequalities in Covid-19 vulnerability and expose their underlying causes. Each chapter focuses on a specific field of international development and examines how the current pandemic amplifies pre-existing challenges and grievances.

We wrote this book having in mind the interested but not necessarily specialist reader – we analyse key issues in a simple and comprehensive manner, without resorting to technical language or assuming prior knowledge of the field. We expect that the book will be particularly helpful for students in economics, political science, anthropology, international relations, geography and international development, who wish to familiarise themselves with the multifaceted implications of the Covid-19 pandemic for developing countries. In addition, we anticipate that the book will be of interest to researchers and practitioners who want to learn about how the current coronavirus pandemic fundamentally reshapes existing debates and processes in international development.

The Hague, The Netherlands   Elissaios Papyrakis

Acknowledgements

We are grateful to many of our colleagues and friends, who generously provided their feedback and ideas (and helped us shape the book in its current form). We have been very fortunate in having Springer's generous support (and patience) towards this research endeavour – we are especially grateful to Ambrose Berkumans, Margaret Deignan, Karthika Menon, Metilda Nancy Marie Rayan and Hermine Vloemans for their continuous guidance and feedback across various stages of the process. We are also much indebted to our family and friends, who were the first ones to read parts of the book and provide their honest opinion on the covered material.

We wish to dedicate this book to our students who often do not realise how much we ourselves learn from them; they inspire us to think outside the box and they are our fellow companions in our lifelong journey of learning. We have much admiration for them. Their studies have been much disrupted by the Covid-19 pandemic and their social interaction has been largely confined to the virtual space. For almost all of them, this has been a very difficult and challenging year; they had to maintain focus on their studies and assignments, despite worrying about their affected family members and friends at home.

Contents

1 **Covid-19 and International Development: Impacts, Drivers and Responses** 1
Elissaios Papyrakis

2 **Reforming the International Financial and Fiscal System for Better COVID-19 and Post-pandemic Crisis Responsiveness** 9
Rolph van der Hoeven and Rob Vos

3 **COVID-19 and the Threat to Globalization: An Optimistic Note** 29
Sylvanus Kwaku Afesorgbor, Peter A. G. van Bergeijk, and Binyam Afewerk Demena

4 **Experiences of Eritrean and Ethiopian Migrants During COVID-19 in the Netherlands** 45
Bezawit Fantu, Genet Haile, Yordanos Lassooy Tekle, Sreerekha Sathi, Binyam Afewerk Demena, and Zemzem Shigute

5 **Consequences of the Covid-19 Pandemic for Economic Inequality** 59
Syed Mansoob Murshed

6 **The Short-Term Impact of COVID-19 on Labour Market Outcomes: Comparative Systematic Evidence** 71
Binyam Afewerk Demena, Andrea Floridi, and Natascha Wagner

7 **Covid-19 and the Informal Sector** 89
Michele Romanello

8 **Indirect Health Effects Due to COVID-19: An Exploration of Potential Economic Costs for Developing Countries** 103
Natascha Wagner

9	**Effects of COVID-19 on Education and Schools' Reopening in Latin America**	119
	Georgina M. Gómez and G. J. Andrés Uzín P.	
10	**Indigenous People, Extractive Imperative and Covid-19 in the Amazon**	137
	Murat Arsel and Lorenzo Pellegrini	
11	**Covid-19 and Climate Change**...............................	147
	Agni Kalfagianni and Elissaios Papyrakis	
12	**Covid-19 and Water**	157
	Farhad Mukhtarov, Elissaios Papyrakis, and Matthias Rieger	

Index... 175

Chapter 1
Covid-19 and International Development: Impacts, Drivers and Responses

Elissaios Papyrakis

Abstract The current coronavirus pandemic fundamentally reshapes existing debates and processes in international development. The unprecedented (and rapidly evolving) crisis is generating a number of substantial challenges for developing economies. Governments in low-income nations often find it extremely hard to cope with the increased demand for health services, make prompt decisions and put them into action, protect vulnerable segments of society and offer immediate relief to affected economic sectors. The current pandemic influences several development outcomes (in the domains of poverty/inequality, health, education, migration, formal/informal employment, (de)globalisation, the extractive sector, climate change, water and the global financial system).

1.1 Introduction

Since the first known human infections in Wuhan, China in late 2019, the coronavirus SARS-CoV-2 (responsible for the Covid-19 disease and current pandemic) has rapidly spread to almost every single country around the world. There are currently more than three million deaths attributed to Covid-19 (and close to 140 million infections – although the accuracy of these figures depends largely on the testing practices of individual countries). Many developing countries have experienced a very large (cumulative) number of Covid-19 related deaths; Brazil, Mexico and India, for instance, have reported more than 365,000, 210,000 and 170,000 coronavirus-related deaths (as of April 2021; detailed information can be found at the WHO website: https://covid19.who.int). The limited institutional preparedness and poor health infrastructure to deal with the rapid spread of the virus was much

E. Papyrakis (✉)
International Institute of Social Studies (ISS), Erasmus University Rotterdam, The Hague, The Netherlands
e-mail: papyrakis@iss.nl

© The Author(s), under exclusive license to Springer Nature Switzerland AG 2022
E. Papyrakis (ed.), *COVID-19 and International Development*,
https://doi.org/10.1007/978-3-030-82339-9_1

encapsulated in numerous pictures of dead bodies lining the streets of Guayaquil (Ecuador's biggest city) in early 2020 (Benitez et al., 2020).

The effects of the pandemic extend beyond the excess mortality attributed to Covid-19 infections. Prolonged lock-downs, squeezed public budgets, disruption in supply chains and delayed health care for other medical (and mental) problems and diseases have all exacerbated a large number of existing developmental challenges (Oldekop et al., 2020). Lack of institutional preparedness and budget constraints, inadequate infrastructure and limited access to information and public resources constitute a toxic mix that amplifies both health and economic vulnerability (as well as reinforce existing patterns of inequalities across and within countries). Many of the global poor work in the informal economy and are, hence, especially vulnerable to restrictions in mobility (and have found themselves with hardly any means to help buffer the economic shocks and income loss; see Narula, 2020). The combination of the pandemic and widespread poverty has aggravated food shortages and heightened economic insecurity for vulnerable individuals and communities.

The current coronavirus pandemic fundamentally reshapes existing debates and processes in international development (Loayza & Pennings, 2020). Governments in low-income nations often find it extremely hard to cope with the increased demand for health services, make prompt decisions and put them into action, protect vulnerable segments of the population and offer immediate relief to affected economic sectors (Gerard et al., 2020). While all of us battle the pandemic at home, it is equally important to oppose isolationism and avoid losing sight of what happens in other (more vulnerable) parts of the world. Vaccine nationalism not only imposes an unfair disadvantage on developing nations (which often are unable to afford preferential agreements with pharmaceutical multinationals) but also fails to prevent the further mutation of the virus and, hence, protect global health (Gastrow & Lawrance, 2021). There is a need for an in-depth critical discussion on how the current pandemic negatively affects multiple development outcomes (e.g. in the domains of poverty/inequality, health, education, migration, informal employment, (de)globalisation, the extractive sector, climate change, and the global financial system) and, in effect, reverses earlier development gains for the global poor.

1.2 Structure and Intellectual Contribution of the Book

While the Covid-19 pandemic has been affecting all countries around the world (both directly through its health impacts and indirectly through its negative income and trade shocks), there are a number of associated threats that are more pervasive and pronounced for vulnerable low-income groups of individuals and countries. The objective of the book (and, more generally, the common purpose of its eleven complementary chapters that follow) has been to highlight these inequalities in vulnerability and identify their underlying causes. It aims at presenting the multifaceted short and long-term impacts of the pandemic on developing economies, vulnerable communities, and international markets and relations more broadly.

Each chapter focuses on a specific field of international development and examines how the current pandemic amplifies pre-existing challenges and grievances.

Chapter 2 discusses the fundamental weaknesses of the international financial and fiscal system in responding to the challenges of the Covid-19 pandemic. While developed economies can resort to significant fiscal and monetary expansion (at very low cost) to protect livelihoods, developing nations face severe liquidity constraints to do the same for their citizens. Additional financial support for low-income countries provided by the IMF and the World Bank has fallen well short of funding needs and is fraught with policy conditionality working against speedy economic recovery. Donor governments also did not step up to the plate as flows of official development assistance (ODA) fell amidst the pandemic. This is severely inhibiting not only the protection of livelihoods in developing countries, but also their capacity to save lives threatened by the pandemic as their health systems remain severely underfunded and means are lacking to secure access to vaccines. The chapter underpins several reform proposals to structurally increase international financial response capacity to deal better with future shocks and crises, including a. setting up proper mechanisms for international tax coordination and sovereign debt management, b. reforming policy conditionality linked to international development finance, and c. issuance of true international liquidity (in the form of IMF special drawing rights) to leverage vast amounts of additional development finance, not just to respond to the economic consequences of COVID-19 but also to meet the additional investment finance requirements for achieving the Sustainable Development Goals.

Chapter 3 proceeds to analyse the potential effects of the pandemic on globalisation through three lenses: namely, the economic, social and political aspects of our global interconnected societies. Across all three dimensions, the pandemic has reduced the extent of globalisation by diminishing international trade volumes, constraining tourism opportunities and international mobility, and prompting in some cases isolationist political responses. However, any deglobalisation process is likely to be relatively modest with evidence already pointing to a quick recovery and reversal to the pre-Covid-19 situation. Earlier pessimistic scenarios pointing to a repetition of a drastic deglobalisation similar to the one of the 1930s seem to be unfounded. To a large extent, the pandemic even highlighted the importance and need for global concerted action and coordination for problems that transcend national borders.

One of the most visible aspects of globalisation has been the increasing movement of people across national borders; in recent decades, thousands of migrants travelled very far, crossing multiple borders in search of safety and a better future (and, in this way, contributed to the multi-ethnic and multicultural mosaics of our modern societies). Chapter 4 provides an in-depth analysis and reflection on how the current pandemic has impacted the lives of migrants by probing into several health and socio-economic effects. The research is based on 18 interviews with representatives of Ethiopian and Eritrean migrant communities and organisations in the Netherlands. It becomes apparent that migrants are particularly vulnerable to the negative shocks and impacts of the pandemic. Their employment in temporary and precarious jobs, large distance to family members and friends still based in their countries of origin and uncertainty about the future all magnify the coronavirus-related challenges these communities face. Language barriers prevent many migrants to fully access important information on Covid-19 prevention and treatment (and more broadly about health and social benefits). The chapter also explores in detail issues of gender and discusses how the gender-based division of labour within migrant households disadvantages disproportionately women (and more so during the times of the pandemic).

Chapter 5 discusses the consequences of the Covid-19 pandemic for economic inequality, both within and between nations. Typically, negative economic shocks and severe recessions are associated with rises in income inequality (given the higher vulnerability of the poor). This is both due to the disproportionate effect of economic contractions on unskilled labour and informal employment, as well as the limited access of poorer segments of the population to health services and information. In addition, the high degree of labour substitutability (as a result of technological advancements in automation and artificial intelligence) keeps wages suppressed, in contrast to earlier pandemics that induced labour scarcities in relation to other production factors. Worldwide, the relatively mild impact of the pandemic on China's economy will likely increase the population-weighted global inequality. The implications are likely to extend beyond the immediate economic realm; excessive inequality will likely undermine liberal democratic governance and increase populism both in developed and developing economies.

Chapters 6 and 7 concentrate their attention to the Covid-19 impacts on the formal and informal labour markets. Chapter 6 first discusses the possible repercussion for formal employment. The coronavirus pandemic resulted in a dramatic contraction of global economic activity, with potentially devastating effects for the labour market. Such labour effects can be the result of reduced labour productivity (due to sickness) or diminished labour demand by firms due to heightened economic uncertainty. The chapter provides a systematic review of the recent literature and uses meta-analysis techniques (based on 2429 existing estimates from studies focusing both on developed and developing economies) to evaluate the short-term impact of the pandemic on formal employment. It finds no meaningful impacts on earnings, hours worked and (un)employment both for developed and developing economies. While these results present a more optimistic picture that one might have anticipated, there is a risk that some of the Covid-19 labour impacts may materialise with a time lag in the near future.

Chapter 7 deals with the implications of the Covid-19 pandemic for those employed in the informal sector. Typically, informal employment is a much larger source of income for developing countries; in many parts of the developing world, the majority of individuals find employment in the informal sector, which is characterised by precarious working conditions, mediocre income levels and lack of social benefits and support. The current pandemic, similar to other earlier major health crises and shocks, intensifies the precariousness of working conditions for those in the informal economy. Adhering to strict health regulations (e.g. in the form of social distancing) implies loss of income for most informal workers. Many decide, nevertheless, to ignore Covid-19 rules in order to secure some minimum income for their survival, at the risk of infecting themselves and their families. For the vast majority of informal workers, who are characterised by low educational attainments and limited connection to formal employment networks, the transition to formal employment is almost impossible (especially during periods of severe economic contraction and higher unemployment rates). Women, younger individuals and workers from ethnic minorities are particularly affected by the Covid-19 pandemic and its impacts on informal employment.

Chapters 8 and 9 focus on the health and educational impacts of the Covid-19 pandemic. Chapter 8 pays attention to the indirect Covid-19 related deaths, as a result of the pandemic worsening health conditions for young children and prospective mothers, as well as HIV, tuberculosis and malaria patients. The economic costs of these health impacts, together with the cost of delayed health care, can have substantial repercussions for the economic growth of many developing countries. The economic cost of Covid-19 associated reductions in regular hospital discharges can be close to a 1% GDP loss for Sub-Saharan economies (as in the case of Ghana and Sierra Leone). For this reason, wealthier economies and international organisations need to provide generous financial support towards developing countries that will both directly reduce the rate of Covid-19 mutations and infections, as well provide uninterrupted health services for pre-existing health problems and diseases.

Chapter 9 probes into the educational effects of the pandemic, paying particular attention to the specificities and institutional weaknesses of developing economies. It examines in detail the disruption of recurrent (or prolonged) lockdowns and social distancing to educational services. This disruption is not a mere inconvenience but can also deactivate learning engagement and result in a loss of already acquired knowledge – it can also generate a significant loss of interest in education, which subsequently increases school absenteeism and desertion. In the longer term, this disruption can bring about reduced employment opportunities and diminished social cohesion. The authors include an in-depth case study based on data from Latin America, highlighting the unpreparedness of many countries in adopting digital technologies and switching to online education. This digital gap also takes place within countries, with vulnerable communities and poorer households lacking access to new technologies and skills. As a result of this, the pandemic is currently reversing prior progress in the domains of educational quality and poverty alleviation.

The last three chapters of the book link the ongoing pandemic with natural resource management across three dimensions: the extractive industry, climate change and water. Chapter 10 discusses how the presence of extractive industries influences the resilience of indigenous communities to external health shocks (as in the case of the Covid-19 pandemic). The study pays particular attention to the challenges posed by Covid-19 for indigenous populations in the Amazon. These are communities characterised by high poverty levels and very limited access to health services due to their remoteness. The arrival of extractive industry and external employment has several negative repercussions for the ability of these communities to cope with Covid-19 (and more broadly with external threats). It has deprived them of the ability to regulate human movement and limit contact with outsiders. At the same time, exposure to extractive activities (and the market economy more generally) gradually erodes ancestral knowledge on traditional medicine and health practices (which is often the sole protection against diseases given the widespread lack of access to alternative health facilities). This is the outcome of a general deep-rooted 'extractive imperative' that promotes mineral extractions as a necessary component of national development, even when this comes at the expense of indigenous communities and their rights.

Chapter 11 focuses on the Covid-19 pandemic and its implications for global warming and associated climate policies. As a result of the current economic downturn and restrictions in mobility, there has been a significant (but likely temporary) drop in carbon emissions. While environmentalists and climate scholars have welcomed this decline in greenhouse gasses, there is a need to devise a long-term strategy that ensures a more sustained transformation towards a relatively carbon-free economy. Along these lines, governments and international organisations need to ensure that there is no reduction in funding towards climate change mitigation and adaptation (either as a result of budget cuts or diversion of funds towards other objectives). Responsibility naturally also lies with each individual. The pandemic offers a unique opportunity to reflect on our current unsustainable lifestyles and incentivise behavioural changes that improve our collective well-being and safeguard climate stability (and protection of the natural environment more broadly).

Chapter 12 probes into the linkages between the Covid-19 pandemic and the water sector, paying particular attention to how the ongoing health crisis exacerbates pre-existing problems and challenges. Limited access to clean water and sanitation has been a chronic problem in many developing countries, but also for vulnerable communities in wealthier nations. The pandemic and limited access to water can lock vulnerable individuals in a reinforcing vicious circle, where water scarcity leads to higher infection rates, poor health, constrained abilities to earn income and a further reduced affordability of water services. At the same time, poor water governance and other combined factors have increased food import dependency in many parts of the world; the current disruption by the pandemic to global supply chains and trade has introduced a heightened risk of food shortages for these countries. For these reasons, both more macro-scale policies are urgently needed that safeguard the uninterrupted provision of funds for the water sector (in the form of public/private investment, international aid and support for water utilities), as well as more micro-scale interventions that incentivise positive behavioural changes regarding water use and sanitation.

1.3 Some Key Messages

While each of the book chapters has a distinctive thematic focus, jointly they allow us to draw some common key messages regarding the Covid-19 impacts on developing countries and appropriate responses.

1 Covid-19 and International Development: Impacts, Drivers and Responses 7

Message 1. *Many of the global poor work in the informal economy and are, hence, especially vulnerable to restrictions in mobility.* The current pandemic has intensified the precariousness of their working conditions and severely reduced their income. Their vulnerability is further enhanced by their inability to access formal relief funding and other related social protection benefits. Formal employment, on the other hand, appears to be much more resilient. (Chaps. 5, 6 and 7).

Message 2. *Vulnerability is further enhanced by unequal access to information.* Poorer communities and individuals often face restricted access to vital information, both in relation to Covid-19 prevention and treatment, as well as with respect to the provision of socio-economic support in periods of financial distress (Chaps. 4 and 5).

Message 3. *Developing countries have fewer means to protect their economies from the Covid-19 shock.* They are unable to resort to fiscal and monetary expansion in order to stimulate their economies and minimise the health repercussions of the pandemic (Chap. 2).

Message 4. *Linked to the above, the pandemic exacerbates existing inequalities, hinders the provision of public goods (locally, nationally and globally) and hence slows down (or even reserves) progress towards the Sustainable Development Goals (SDGs).* Several intertwined factors (the economic downturn, increases in unemployment, health impacts, rise in poverty, unequal distribution of impacts, informational asymmetries etc) make it increasingly more difficult for developing countries to meet the SDGs (Chaps. 2, 3, 5, 8, 9, 10, 11 and 12).

Message 5. *The pandemic has a negative effect on globalisation, which is likely though to be of a temporary nature.* The pandemic coincides with reduced international trade volumes, diminished tourism opportunities and constrained international mobility, although there is already evidence pointing to a quick recovery. In several cases, there is evidence of isolationist political responses and lack of international solidarity (Chaps. 2 and 3).

Message 6. *Worsening health conditions and delayed health care impose a substantial economic cost to developing countries.* Limited access to water is an additional chronic problem in many developing nations that reinforces a vicious circle of higher infection rates, poor health, constrained abilities to earn income and reduced affordability of water services (Chaps. 8 and 12).

Message 7. *The pandemic generates a substantial disruption to educational services in developing countries.* This relates to both the institutional unpreparedness in switching to online education, as well as the overall lack of access to digital technologies and skills for low-income households (Chaps. 9 and 11).

Message 8. *There is a gendered dimension of the Covid-19 impacts.* The gender-based division of labour disadvantages disproportionately women (who often bear the entire responsibility of schooling kids and taking care of other family members during the lockdown). Impacts on informal employment also particularly affect women (and other vulnerable groups) who find it much harder to participate in paid work (Chaps. 4 and 7).

Message 9. *Indigenous communities and ethnic minority groups are particularly vulnerable to the health and socio-economic impacts of the pandemic.* Exposure to the market economy and external employment limits the ability of indigenous populations to regulate human movement and preserve ancestral knowledge on health practices (Chap. 10). Ethnic communities within developed countries also face severe limitations in accessing relevant information and job opportunities during the pandemic (Chaps. 4 and 7).

Message 10. *Interest in environmental issues may diminish as a result of the pandemic.* There is a risk of reduction in funding for climate change mitigation and adaptation, as well as for for water security (as a result of budget constraints and diversion of funds towards other objectives). Limited access to water and accelerated global warming, however, make the occurrence of future pandemics more likely and can intensify their impact on vulnerable individuals (Chaps. 11 and 12).

References

Beniitez, M. A., Velasco, C., Sequeira, A. R., Henriquez, J., Menezes, F. M., & Paolucci, F. (2020). Responses to COVID-19 in five Latin American countries. *Health Policy and Technology, 9*(4), 525–559.

Gastrow, C., & Lawrance, B. N. (2021). Vaccine nationalism and the future of research in Africa. *African Studies Review, 64*(1), 1–4.

Gerard, F., Imbert, C., & Orkin, K. (2020). Social protection response to the COVID-19 crisis: Options for developing countries. *Oxford Review of Economic Policy, 36*(S1), S281–S296. https://doi.org/10.1093/oxrep/graa026

Loayza, N., & Pennings, S. (2020). *Macroeconomic policy in the time of covid-19: A primer for developing countries* (World Bank research & policy briefs 28). World Bank.

Narula, R. (2020). Policy opportunities and challenges from the COVID-19 pandemic for economies with large informal sectors. *Journal of International Business Policy, 3*(3), 302–310.

Oldekop, J. A., Horner, R., Hulme, D., Adhikari, R., Agarwal, B., et al. (2020). COVID-19 and the case for global development. *World Development, 134*, 105044.

Chapter 2
Reforming the International Financial and Fiscal System for Better COVID-19 and Post-pandemic Crisis Responsiveness

Rolph van der Hoeven and Rob Vos

Abstract The global economic crisis provoked by the COVID-19 pandemic disproportionally hurt developing countries, increasing poverty, food insecurity, and income inequality. Richer nations cushioned their economies from the worst impacts with unprecedented massive fiscal and financial support programmes. Developing countries lacked such capacity and received feeble multilateral contingency financing, symptomizing the fundamental flaws in the international financial and fiscal system (IFFS). Four reforms will make the IFFS better suited to serve sustainable development: (a) an equitable international tax coordination mechanism; (b) a multilaterally backed sovereign debt workout mechanism; (c) overhauling policy conditionality associated with development finance; and (d) increasing Special Drawing Rights to be leveraged for development finance.

2.1 Introduction

The global economic crisis provoked by the COVID-19 pandemic once more has painfully revealed fundamental flaws in the international financial and fiscal system (IFFS). It failed to provide adequate crisis response, especially to the low-income countries which have been hard hit by the global economic repercussions of the pandemic. Even though the spread of the pandemic was less pervasive in affecting

---

R. van der Hoeven (✉)
UN Committee for Development Policy, New York, USA

International Institute of Social Studies (ISS), Erasmus University Rotterdam,
The Hague, The Netherlands
e-mail: rolph@vanderhoeven.ch

R. Vos
International Institute of Social Studies (ISS), Erasmus University Rotterdam,
The Hague, The Netherlands

Markets, Trade and Institutions Division (MTID), International Food Policy Research Institute (IFPRI), Washington, DC, USA

© The Author(s), under exclusive license to Springer Nature Switzerland AG 2022
E. Papyrakis (ed.), *COVID-19 and International Development*,
https://doi.org/10.1007/978-3-030-82339-9_2

health conditions in many low-income countries, they were hit disproportionally hard by the spill-over effects from the economic crisis in major economies, especially those of Europe and the United States, while possessing few financial means to mitigate the impact on livelihoods of their populations.

In this chapter, we assess the differential impact of the pandemic on livelihoods around the world. Specifically, Sect. 2.2 describes how the COVID-19 crisis has set off a deep global recession with potentially lasting setbacks in terms of human development, increasing poverty, income inequality and food insecurity. Some of the poorest countries, as those in Africa, are among the hardest hit economically even though being, thus far, less hurt by the spread of the pandemic itself. While high-income countries have taken unprecedented measures to mitigate impacts on livelihoods, poor nations' responses have been muted lacking financial space.

Section 2.3 then documents how the economic consequences of the pandemic have once more lay barren the weaknesses of the current IFFS, which fails to allocate sufficient contingency financing where it is needed the most. Where high-income countries could engage in massive and almost costless fiscal and monetary expansion, developing nations ran up debts to severe distress levels and faced liquidity shortages for undertaking even the smallest of deficit financing. This lack of fiscal space and access to finance has its origins in the weaknesses of the IFFS, including the pervasive biases in internationally poorly-coordinated tax systems causing tax base erosion and profit shifting, the inadequacy of international contingency financing mechanisms, the lack of appropriate sovereign debt management mechanisms, and the absence of a truly international currency that could serve both as a multilateral source of liquidity provisioning during crises and a basis for leveraging development finance to build resilience against the impacts of future crises.

Section 2.4 discusses options on how these weaknesses in the IFFS can be addressed in the short and medium run, including through reform of international tax coordination mechanisms, putting in place a multilaterally backed sovereign debt workout mechanism, reform of policy conditionality attached to contingency financing, and issuance of additional, truly international liquidity in the form of Special Drawing Rights both to provide additional contingency finance and to leverage new development finance. With such reforms already in place, the pandemic response would have provided a fairer level playing field for emerging and developing countries and have mitigated the pandemic's worst economic consequences. Beyond the pandemic, these reforms will aid the recovery and refocus international development finance towards the internationally agreed Sustainable Development Goals. As discussed in the concluding section, important political hurdles will have to be overcome to enact those reforms in the present-day context of withering multilateralism. It is of interest to note here, that since the time of writing (April 2021), the international community has taken several steps towards the fulfilment of these recommendations. First, the Group of Twenty (G20) of major economies agreed to establish a common 25% corporate tax rate in a concerted effort to discourage profit shifting by multinational companies, a phenomenon that is currently eroding the tax base of many countries. And, second, approval by the International Monetary Fund (IMF) of the issuance of US$650 billion in Special Drawing Rights (SDRs).

Although, a proper allocation mechanisms for use of the additional international liquidity is yet to be agreed upon, it is an important step in the right direction. It shows that reform is possible, though there is still a long way to go, as this chapter makes clear.

2.2 Economic, Social and Nutritional Effects of the COVID-19 Crisis

2.2.1 Economic and Social Effects

More than half of the world population has been, still is or is again under some form of social distancing regime designed to contain the health crisis posed by COVID-19. Business activity has fallen sharply because of a combination of policy action and personal responses designed to reduce risk of contracting the virus, with personal action probably more important than policy in reducing economic activity (Goolsbee and Syverson, 2020). Disruptions in production and income losses underpinned the combined supply and demand shock provoking a global recession much deeper than that of the global financial crisis of 2008–2009. The International Monetary Fund (IMF, 2021a, b) projects that the impacts could well be felt for some time to come. It estimates that despite tangible recovery in 2021, global GDP would still be 3.7% below pre-COVID levels by January 2022 (Fig. 2.1). Developing regions are being hit the hardest, with GDP losses in developing Asia (excluding China) estimated at 8% and in Latin America and Sub-Saharan Africa at 6.9% and 6.1%, respectively. Aside from supply disruptions caused by lockdown measures taken by governments in these regions, developing countries are hurt foremost by the transmission of the recession in Europe and the United States through channels of trade, finance and remittances.

A key symptom of the recession has been the substantial loss of working hours throughout the world. The International Labour Organization (ILO, 2021) documents that in 2020 globally, around half of working-hour losses were due to employment loss, while the other half can be attributed to reduced working hours (including workers who remain employed but are not working, see Fig. 2.1). It found further that there was significant variation between regions: employment losses, both as a share of the working-age population and in relation to working-hour losses, were highest in the Americas, and lowest in Europe and Central Asia, where reduced working hours have been extensively supported by job retention schemes. The ILO estimates that, despite the adjustment through reduced working hours, employment losses were nonetheless massive in 2020, with 114 million jobs lost relative to the pre-crisis employment level in 2019. However, this estimate may well understate the full extent of job losses: a model-based scenario analysis suggests global employment decline could have affected as many as 144 million jobs compared with a "no-pandemic" scenario (ILO, 2021).

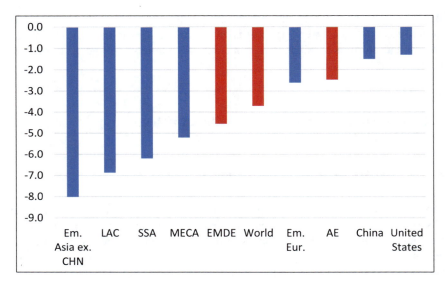

Fig. 2.1 GDP losses (projected levels for January 2022) relative to pre-COVID output levels (percentage change)
Source: IMF World Economic Outlook, January 2021 (IMF, 2021a)
Legend: Em. Asia = ex CHN emerging Asia, excluding China; LAC = Latin America & Caribbean; SSA = Sub-Saharan Africa; MECA = Middle East and Central Asia; EMDE = group total for emerging and developing economies; Em.Eur = emerging Europe; AE = advanced (high-income) economies

In stark contrast with the Great Recession of 2008–2009, the COVID-19 related fall in employment mainly translated in rising inactivity rather than rising unemployment rates: of the mentioned job losses 81 million refer to people shifting into economic inactivity, while 33 million became unemployed. As a result, the global labour participation rate dropped by 2.2% points during 2020 (compared with a just 0.2% points decline between 2008 and 2009). As shown in Fig. 2.2, only high-income countries saw, on average, a greater increase in unemployment than in inactivity (driven largely by labour market adjustment in the United States).

Owing to the massive losses in working hours, workers suffered large reductions in their income from work. The ILO estimates that labour income declined by 8.3% in 2020 relative to 2019 (Fig. 2.3). The greatest labour income loss, amounting to 12.3%, was experienced by lower-middle income countries. While average labour income losses were of similar magnitude in low-, upper-middle- and high-income countries, these averages disguise large disparities across and within these country groupings. When looking by geographic region, it shows that workers in the Americas lost an estimated 10.3% of their labour income, compared with 6.6% for workers in Asia and the Pacific.

Overall, in nominal terms, global labour income declined by about US$3.7 trillion (using 2019 market exchange rates) during 2020, corresponding to 4.4% of global GDP in 2019 (ILO, 2021).

2 Reforming the International Financial and Fiscal System for Better COVID-19... 13

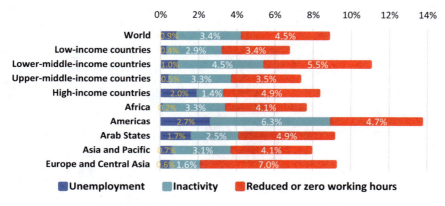

Fig. 2.2 Decomposition of working-hour losses into changes in unemployment, inactivity and reduced working hours (averages for the world and by income group and region in 2020, percentage change)
Source: ILO (2021: Fig. 7).
Note: The overall working-hour loss is decomposed into changes in unemployment, inactivity and reduced or zero working hours. Unemployment plus inactivity equals the total employment loss. Unemployment and inactivity have been transformed into their working hour equivalent using the average working hours per week. The working-hour equivalent of changes in employment, unemployment and inactivity is computed using the estimated average working hours per week, which ranges from 35 to 48 h per week across the income groups and regions.

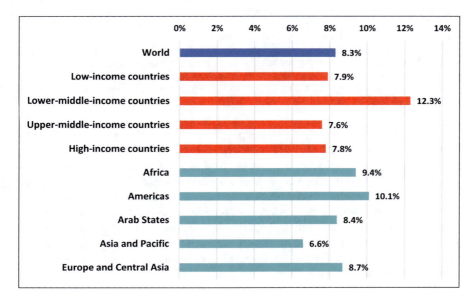

Fig. 2.3 Share of labour income lost due to working-hour losses in 2020, before income-support measures (average for the world and by income group and region, percentage change)
Source: ILO (2021: Fig. 9).
Note: Labour incomes were aggregated across countries using purchasing power parity exchange rates.

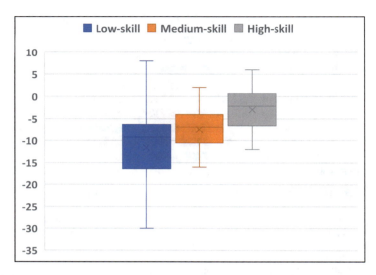

Fig. 2.4 Year-on-year country-level changes in employment, by skill level, second quarter of 2022 (percentage change)
Source: ILO (2021: Fig. B2).
Note: The sample consists of 50 high- and middle-income countries and territories with employment data for the second quarter of 2020 and disaggregated by occupation. The box graph should be read as follows: (a) the vertical line in the middle of the box represents the median value (50th percentile; (b) the lower side of the box (whisker) represents the 25th percentile; (c) the upper side of the box (whisker) represents the 75th percentile; (d) the adjacent lines to the above and below the box represent the highest and lowest values, respectively
Legend: Low-skill = elementary occupations and skilled agricultural workers; Medium-skill = clerical support workers, services and sales workers, craft and related trades workers, plant and machine operators, and assemblers; High-skill = managers and technicians, and associate professionals. The skill level categories are based on ISCO-08. See ILOSTAT for further details on these definitions

Labour income losses caused by the pandemic show stark inequalities between groups of workers (Murshed, 2021). Generally, low- and medium-skilled workers were much more affected by employment and income losses than better skilled workers (Fig. 2.4). This is in part because teleworking more often proved an option for those workers, while social distancing measures hampered executing many lower skilled jobs. Where the capacity to strengthen social safety nets is weak, the labour income losses pushed many households into poverty (forcing them to reduce spending on necessities once savings were used up) and further deepened the recession because of demand fallout. We take a closer look at the poverty and consumption impacts in the next subsection.

2.2.2 Poverty and Food Consumption Effects

Assessing the poverty impact of COVID-19 is no trivial matter. This is so not only because the crisis is still unfolding and available information of its precise socio-economic consequences is incomplete, but also because the channels of influence are multiple and interconnected globally. Several widely cited analyses have used simplistic approaches calculating the projected impact of the global recession on average per capita incomes to estimate poverty impacts, using household survey data available through the World Bank's PovcalNet website (see, for example, the studies by the World Bank in Mahler et al., 2020 and World Bank, 2020a, 2020b; and that of UNU-WIDER by Sumner et al., 2020). A major drawback of this approach is that it assumes that the crisis has had no impact on within-country income distribution and, consequently, that workers across sectors and type of activity were all affected to the same degree.

Laborde et al. (2021) point out that this assumption fails to account for the complexity of the channels of effect and may substantially underestimate the impacts of the pandemic. They use a global general equilibrium model linked to country-specific household models to simulate the impacts of the COVID-19 pandemic on poverty and food security, considering all key transmission channels, including the labour market impacts discussed in the previous sub-section.[1] Beyond the direct effects of the disease on the ability to work, income losses arise from people's desire to avoid catching the disease and their altruistic concerns to avoid infecting other people, and from policy responses designed to reduce the adverse externalities associated with an unmitigated pandemic. Many of the related changes in behaviour and in the functioning of economies are not yet fully understood, while it is also difficult to rely on experience from past events, since no events like the COVID-19 pandemic have occurred on this scale in today's globalized world. Therefore, Laborde et al. (2021) have had to make several assumptions about the responses of economic agents to this unprecedented situation.

They distinguish four drivers of COVID-19 impacts: domestic supply disruptions, global market disruptions, household behavioural responses and policy responses. In a scenario with assumptions based on evidence available by September 2020, Laborde et al. (2021) project a 7% decline in global GDP in 2020 (compared with a scenario without COVID-19) and, consistent with the IMF projections presented earlier, they show that developing countries are being hurt disproportionally through declines in trade and remittance incomes and disruptions in businesses caused by social distancing measures.

Without any significant social and economic mitigation measures, e.g. in the form of a fiscal stimulus and expansion of social safety nets in the global South (scenario assumption, but see also Sect. 2.3 below), the impact on extreme poverty (measured against the PPP$1.90 per person per day international poverty line) is devastating, as shown in Fig. 2.5. The number of poor increases by 20% (almost

---

[1] For details and model assumptions see Laborde et al. (2020, 2021: online appendix).

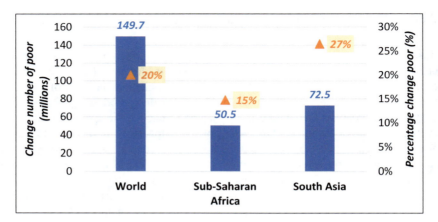

Fig. 2.5 Global and regional poverty impacts of the MIRAGRODEP COVID-19 scenario (September 2020; absolute and percentage change from baseline values)
Source: Laborde et al. (2021: Table 2).

150 million people) with respect to the situation in the absence of COVID-19, affecting urban and rural populations in South Asia the most, where 72.5 million more people would be joining the ranks of the poor (equivalent to a 27% increase in that region). The poverty increase in rural areas is expected to be smaller than that in urban areas, partly because of the lower rate of transmission of the disease and partly because of the robustness of demand and supply for food relative to many other, more vulnerable sectors. The number of poor people in Sub-Saharan Africa is projected to increase by 15% or 50.5 million people.

A decomposition of the poverty impact shows that the estimated increase in the number of poor by 150 million is substantially higher (i.e., about 50 million more) than when assuming a uniform drop in average per capita incomes in each country, as done by the earlier cited studies of the World Bank and UNU-WIDER. It indicates that COVID-19 must have significantly worsened within-country income inequality, as well.

The income and price changes associated with the recession and supply disruptions caused by pandemic are furthermore likely to have resulted in substantial changes in patterns of food consumption, with adverse nutritional consequences. Laborde et al. (2021) project that these will induce shifts in demand away from nutrient-dense foods, such as fruits and vegetables, dairy products and meats, and towards calorie-rich basic staple foods, such as rice, maize and other basic grains, raising concerns about dietary quality and likely increase in micronutrient deficiencies. The dietary shift is (on average) similar in both developed and developing regions.

2.3 Fiscal and Financial Crisis Responses: Save Thyself First?

The strongly integrated world economy facilitated the rapid spread of COVID-19 and its economic repercussions around the world. As we showed in the previous section, poorer nations and more vulnerable households have been hurt disproportionally by the economic consequences of the pandemic. Economic response capacity was very uneven. While rich countries engaged in unprecedented responses in macroeconomic terms to mitigate impacts on livelihoods of their citizens, most of the worst-hit countries lacked such economic response firepower. Multilateral mechanisms should have provided a cushion but proved to be unfit for such purpose.

2.3.1 The Scramble for Access to Vaccines

In Spring 2020, there was certainly a greater awareness than before of a necessary global response including the need to save lives. A search for and the production of reliable vaccines took off with much national government support. Internationally, the World Health Organization (WHO) swiftly moved to set up its ACT[2] Accelerator and the COVAX.[3] The purpose of these mechanisms was to finance the acceleration of the development of COVID-19 vaccines, to secure availability of sufficient doses for all countries, and distribute those doses fairly, beginning with the highest risk groups and spreading to cover the entire global population as soon as possible. However, because of limited and falling domestic resources in developing countries, by April 2021, both the production and distribution of vaccines was still heavily skewed in favour of serving the populations in high-income countries. The COVAX remained heavily underfunded more than a year into the pandemic, and unable to buy the vaccines needed to cover even a fraction of the population of developing countries, as global supplies have been by and large captured by the wealthier nations.

In effect, inequalities in access to vaccines to combat the pandemic have been reflective of how financial responses have worked out. A lagging behind in vaccination rates in developing countries has serious repercussions for global immunity rates as the virus still can mutate and cause again worldwide infections compounding already grave economic consequences.

A further complicating factor is that production of vaccines in poorer countries is hampered by persistence of patent rights allowing pharmaceutical companies a monopoly of production, either by themselves or by companies able to purchase

---

[2] ACT stands for Access to Covid-19 Tools (https://www.who.int/initiatives/act-accelerator)

[3] COVAX stands for COVID-19 Vaccines Global Access Facility, led by the World Health Organization, the Coalition for Epidemic Preparedness Innovations, and Gavi, the Vaccine Alliance. See: https://www.who.int/initiatives/act-accelerator/covax

production licenses. The WTO agreement on Trade-Related Intellectual Property Rights (TRIPS) allows for compulsory licensing, while the Doha declaration on TRIPS and public health contains an explicit clause permitting compulsory licensing essential drugs in case of public health emergencies.

A refusal by rich countries to set in motion existing clauses in international trade rules and underfunding of financial support to purchase and develop vaccines in poorer countries resulted in a highly unequal distribution of vaccines towards these countries. By Spring 2021, 130 countries had not yet administered a single dose to their populations and at current rates of distribution some people in developing countries will not receive a vaccine until 2024 (INET, 2021: 7). A continuing threat of COVID-19 contamination in countries with low vaccination levels increases social and economic uncertainty hampering prospects for economic recovery and progress. Furthermore, it is no coincidence that developing countries, where the availability of vaccines is low, have also less resources to stimulate their economies.

2.3.2 Stark Inequalities in Financial Response Capacity

Governments not only need additional resources to address the health impacts of the COVID-19 crisis and for the development and roll-out of vaccines, but also to finance the costs of the various lockdowns necessary to halt the spreading of the virus as well as to stimulate the economy to make up for the fall in final demand caused by the COVID-19 crisis. Globally, fiscal and monetary stimulus and emergence support measures amounted to US$14 trillion or 13.5% of world GDP. Fiscal support measures summed to almost US$8 trillion (or 8% of world GDP). The government response was vastly larger than that following the 2008–2009 global financial crisis, and likely has prevented a much deeper global recession. Inequalities in the fiscal and monetary response capacity run as deep as the overall response has run high. Figure 2.6 shows the stark differences between high-, middle- and low-income countries. High-income countries provided fiscal stimulus to the tune of 12.5% of GDP on average; this was three times more in relative terms than emerging and other middle-income countries were able to provide and almost ten times more than that provided by governments in low-income countries. In per capita terms, the inequalities run even deeper. UN-DESA (2021) estimates that the stimulus packages per capita by the developed countries has been nearly 580 times bigger in size than those enacted by the UN category of least developed countries (LDCs). Put in perspective, the average income per capita of developed countries is "only" 30 times that of the LDCs.

Lacking sufficient domestic resources to support their economies to pay for health costs and to stimulate their economies developing countries need to look for outside resources. This proved difficult for many developing countries for several reasons.

First, after a period of falling external debt levels supported by the Highly Indebted Poor Country (HIPC) initiative during the 2000s, external debt burdens of

2 Reforming the International Financial and Fiscal System for Better COVID-19...

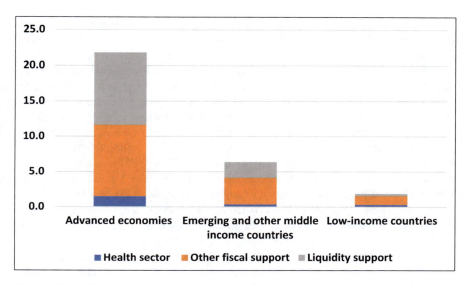

Fig. 2.6 Fiscal and monetary support in response to COVID-19 per January 2021 (% of GDP) *Source*: IMF (2021b), Fiscal Monitor, Database of Country Fiscal Measures in Response to the COVID-19 Pandemic

many low-income developing countries surged again during the 2010s. In 2019, the IMF assessed that half the low-income countries faced high risk of or already were in debt distress; more than double the 2013 share. The average external debt ratio of low-income countries had already increased to 65% of GDP in 2019, up from 47% in 2010. Increased borrowing from private lenders significantly added to the public debt burden in many countries, with the share of private non-guaranteed debt in total external debt stocks of low-income countries increasing from 3.2% in 2010 to 10% in 2019 (Chandraskhar, 2021). During 2020, these high levels of external indebtedness caused substantially more distress with the steep declines in GDP and export earnings owing to the global recession caused by the pandemic.

To alleviate some of the stress, the IMF cancelled US$213.5 million in debt-service obligations for 25 eligible HIPCs during 2020. This debt relief, while welcome, proved far from sufficient, however, to avoid increased debt distress. Likewise, the G20's debt service suspension initiative (DSSI) thus far has provided little debt relief and, in essence, only helped kick the can down the road, as no debt was cancelled, with interest continuing to accrue during the all-too-brief suspension period (Chowdury & Jomo, 2021).

Second, during 2020, developing countries also faced an outflow of capital to developed countries and were hit by the appreciation of the dollar (and depreciation of their own currencies). Gallagher et al. (2021) point to the high degree of uncertainty and an initial lack of coordinated policy responses, which intensified market panic and volatility; this led to the largest outflow of portfolio capital from emerging market and developing economies in history and caused a global shortage of dollar liquidity. At the same time, the external financing needs of most developing countries increased staggeringly, as they saw export earnings collapse with the global demand fallout (and, for many, this was compounded by declines in major commodity prices), while the currency depreciation increased their import bills. As per the above, their financing needs further increased because of the need to combat the health and economic impacts of the pandemic. Despite a general awareness and pledges made by the G20, in practice, the response of IMF and the World Bank was far from commensurate with the magnitude of the crisis (see Afesorgbor et al., 2021). While the IMF indicated it would commit its US$1 trillion lending capacity, as of March 15, 2021, it had provided no more than US$107 billion worth of financial assistance to 85 countries around the world. The World Bank announced a US$160 billion pandemic support program in September 2020. While significant, elements of this program have been criticized for not helping to remove fiscal barriers and lack of attention to accessible healthcare. Only 8 of the 71 COVID-19 health projects (funded by the World Bank) include measures to support low-income people that now face financial barriers to access health services (Oxfam, 2021a).

Third, the additional resources that were made available have come with conditions that in fact made these less effective as a response to the impacts of the pandemic. An analysis of the contents of recent and ongoing IMF agreements by Oxfam (2020) revealed that, between March and September 2020, 76 out of the 91 IMF loans for 81 countries with a total value of US$89 billion had conditionality attached that required recipient governments to slash public expenditure in ways that could result in deep cuts in the funding of public healthcare systems and pension schemes, while requiring economy-wide wage freezes and reducing public sector employment. Nearly one-third of the countries with IMF loans also face surcharges on unpaid interest (amounting to more than US$4 billion) even in the midst of the pandemic, substantially increasing debt-servicing cost (INET, 2021: 10).

In short, the current IFFS proved inadequate as a financial safety net for countries put in extraordinary need because of a pandemic. As a result, it only exacerbated global inequalities given that countries with ample fiscal and monetary resources were able to take unprecedented fiscal and financial support measures to protect the livelihoods of their own populations but without much consideration of dealing with the international repercussions on the livelihoods of the populations in less affluent nations. This apparent lack of international solidarity during the pandemic is further reflected in a projected decline of almost 40% in bilateral official development assistance during 2020 (Development Initiatives, 2021).[4]

---

[4] According to a February 2021 briefing by Development Initiatives (2021), bilateral donors have decreased aid commitments by 36% between 2019 and 2020 (over the same January to November period). Of the thirteen bilateral donors considered in this analysis (covering 97% of 2020 bilateral commitments by value), seven have seen total ODA commitments fall, with four seeing falls by 40% or more.

2.4 Changes Needed to Address the Weaknesses of the Current International Financial and Fiscal System (IFFS)

The COVID-19 crisis revealed that the current IFFS is unfit to provide adequate emergency and development finance to countries and populations most in need, thereby exacerbating global inequality and impeding the attainment of the Sustainable Development Goals by 2030 (see also Mukhtarov et al., 2021). Reform of the IFFS is thus an imperative of the first order. The global nature of the pandemic might give impetus to such reform, but the political obstacles remain vast. Without attempting to be comprehensive, we focus on four key reform proposals: (i) a much stronger international tax coordination, including harmonizing higher corporate tax rates (especially on profits of globally operating firms) and the reduction of Base Erosion and Profit Shifting (BEPS) in developing countries; (ii) sovereign debt restructuring and relief; (iii) reform of policy conditionality attached to lending by the international financial institutions (IFIs); and (iv) an increase in international liquidity by issuing additional Special Drawing Rights (SDRs), including for leveraging additional developmental finance.

2.4.1 International Tax Reform

Lower taxes and various forms of tax avoidance have left governments with less resources to face important priorities in the wake of the COVID-19 pandemic. Tax avoidance diverts 40% of foreign profits to tax havens. This "profit shifting" causes estimated government revenue losses between US$500 billion and US$600 billion per year (FACTI, 2021). Tax evasion by wealthy individuals and illicit financial transactions add to the substantial amounts of forgone government revenue.[5] Illicit flows are expected to have increased during the pandemic.

The Independent Commission for the Reform of International Taxation (ICRIT) has proposed the following changes in the international tax system: a higher corporate tax rate to large corporations in oligopolistic markets allowing them to earn excess rates of return; a minimum effective corporate tax rate of 25% worldwide to stop base erosion and profit shifting; progressive digital services taxes on the economic rents captured by multinational firms; country-by-country reporting for all corporations benefitting from state support; publication of data on offshore wealth

---

[5] The High-Level Panel for Financial Accountability, Transparency, and Integrity (FACTI) estimates that about US$7 trillion in private asset holdings is kept hidden from tax collectors in tax-haven countries; 10% of world GDP may be held in offshore financial assets, while an additional US$20 billion to US$40 billion is estimated to be paid in the form of bribes on investment deals. Furthermore, revealed money-laundering transaction by criminals are estimated to represent 2.7% of global GDP. Estimates drawn from the FACTI interim report (FACTI Panel, 2021).

to enable all jurisdictions to adopt effective progressive wealth taxes on their residents and to better monitor effective income tax rates on highest income taxpayers (ICRIT, 2020). These measures would greatly increase the fiscal space in low and in middle income countries.

2.4.2 Sovereign Debt Restructuring and Relief

Twenty years after the start of the HIPC initiative for debt relief and restructuring, many countries are still in a precarious debt position, as analysed in Sect. 2.3. The Commission on Global Economic Transformation (led by Nobel Prize-winning economists Joseph Stiglitz and Michael Spence) observed that while attention to poor countries' needs for debt relief and restructuring regained some traction in 2020 when the pandemic broke out, this soon fizzled out: 'in the beginning of the pandemic there was an agreement among the G-20 for a moratorium on servicing of the debt for the poorest countries for their official (bilateral) debt, called the Debt Service Suspension Initiative (DSSI), as mentioned earlier. The hope was that others would join, the private sector in particular. But they did not. The lack of comprehensive participation has a devastating effect: those who might be willing to join are hesitant to do so, as they see the net beneficiary not being the poor people in the poor country, but the recalcitrant creditors' (INET, 2021: 11).

The international community should create better conditions for sovereign debt restructuring. The dire situation caused by the COVID-19 pandemic gives all the reasons to accept the principle of force majeure: countries should not be forced to pay back what they cannot afford. In this sense, Gallagher et al. (2021) propose the creation of an appropriate Sovereign Debt Restructuring Regime, building on earlier proposals made in the aftermath of the global financial crisis (see e.g., Herman et al., 2010). Although existing mechanisms to renegotiate sovereign debts with private creditors have improved, they are still far from adequate because of the multiplicity of debt contracts, some of which are not subject to collective action clauses. A global institutional mechanism to renegotiate sovereign debts should therefore be put in place as soon as possible. Many developing countries were already close to external debt insolvency due, in part, to the recent surge in private external borrowing, as noted above. The massive capital flight and exchange rate depreciation that took place during 2020 has compounded the developing-country debt distress, increasing the likelihood of default, and making the need for orderly sovereign debt workouts the more urgent, not only to bailout debt-distressed countries, but also to safeguard global financial stability.

2.4.3 Reform of Policy Conditionality Attached to Lending by IFIs

In Sect. 2.3, we noted the stark disparities across countries at different levels of development in fiscal support in response to the COVID-19 crisis. The IMF plays a large role in the macroeconomic policies undertaken by borrowing developing countries, especially those that are facing balance-of-payments problems and turn to it for advice and support. The Commission on Global Transformation notes with satisfaction that the IMF leadership has actively supported the large multi-year fiscal stimulus for COVID-19 recovery enacted by the United States and most European countries. The IMF has further recognized the need for enhanced public spending by developing-country governments, also those facing debt distress. Unfortunately, as also noted in Sect. 2.3, in practice, the IMF has continued to provide pro-cyclical policy advice to borrowing nations, as reflected in the policy conditionality attached to its loans, asking for fiscal restraint rather than deficit spending when economies are in recession. Between October 2020 and March 2021, the IMF approved an additional US$18.6 billion in new loans to 16 countries, raising the total amount of lending approved during the pandemic to US$107 billion. Nearly all (93%) of the additional funding was allocated to alleviate fiscal stress in Latin America and the Caribbean and only a paltry 3% went to Sub-Saharan Africa, 2% to countries in Asia and the Pacific, and 2% to countries in North Africa, the Middle East and Central Asia. Out of the 18 new loans, 17 loans called for fiscal austerity by recipient countries (Oxfam, 2021b). The Oxfam report further observes that the IMF has introduced flexibility clauses in some non-emergency loans, allowing countries to increase social spending in case the pandemic worsens. Agreements further stress the importance of protecting social spending, also where fiscal consolidation is required. However, the language in the Letters of Agreement is often vague regarding this flexibility, while explicit and precise regarding targets for fiscal consolidation and spending cuts. For instance, where the IMF encourages governments to protect social spending, it also advises governments in the same documents to roll back pandemic-related social spending as soon as "the crisis abates". This is worrying considering that most countries were extremely unprepared to face the crisis with severely insufficient social spending. The emphasis on the need for fiscal restraint by governments receiving financial support from the IMF further points at the inadequacy of the contingency financing to provide the fiscal space needed to mitigate the worst impacts of the pandemic on livelihoods. In this sense, the practice of IMF's policy conditionality still looks very much alike it was before the pandemic, as analyzed by Gallagher and Carlin (2020) upon reviewing a pre-pandemic set of IMF loans.

The specifics of each IMF program should therefore be much better scrutinized before being presented to the board of Directors; any policy conditionality attached to lending policies by IFIs should more consistently be based on principles of supporting a countercyclical macroeconomic policy stance by recipient countries. In addition, short-term lending to face off balance-of-payments should be aligned with adequate long-term development finance in support of achievement of the Sustainable Development Goals.

2.4.4 An Increase in Special Drawing Rights (SDRs) with Special Usage for Developing Countries

At the time of writing, the IMF was expected to approve the issuance of US$650 billion in new Special Drawing Rights (SDRs), and effectively did so in June 2021. The advantage of creating international liquidity in this way is that it is essentially costless. Earlier fears that this could be inflationary are not relevant in the current global economic conditions, especially as it is dwarfed by the monetary expansion in rich countries. This increase provides developing countries immediately with an increase in their reserves and enable them to engage in much-needed public expenditure with less concern for the effects on the external balance; it could also provide some means of repayment for countries with pressing external debt problems. This impact of the new SDR creation is not automatic, however, since SDR allocations are currently made in line with IMF quota (which in turn are linked to voting rights). Most of the new SDRs therefore will be made available to richer countries who are less in need of balance of payments finance and hence much of the new SDRs could remain unused reserves within the IMF. Any unused SDR reserves could be made available to developing countries in several ways (INET, 2021: 9; United Nations, 2021). First, it could be decided to use a part for writing off or reducing the external public debts of poor countries. Second, SDRs could be given or lent to specific countries with high balance-of-payments stress. Third, as has been proposed in the past (see, e.g., Haan, 1971; United Nations, 2012; Ocampo, 2015; Vos, 2017), a portion of unused SDRs not needed as a reasonable reserve buffer could be leveraged for development finance (through issuance of international bonds backed by those unused SDRs).

2.5 Conclusions

The COVID-19 pandemic and its social and economic outfall has led to a steep decline in economic activity, as other pandemics did (van Bergeijk, 2021). But people and countries have not been equally affected. In high-income countries in Europe and North America, the pandemic has increased inequality, as highly trained workers and capital owners were much less affected that other groups. Developing countries, and especially the vulnerable segments of their populations, were disproportionally hit, resulting in more poverty, greater food insecurity and worsening nutrition. As developed countries had greater fiscal space, they could inject ample resources into necessary health measures, expansion of social safety nets and economic stimulus, something which most developing countries were not able to do. Thus far, the multilateral system failed to come to their financial rescue, owing to waning of multilateralism itself (van der Hoeven, 2020) and, importantly, because of fundamental shortcomings in the IFFS to serve as an international financial and social safety net in a global crisis. Proposals to reform the existing IFFS prop up with each crisis, but, as this chapter has made clear, if the key changes described in Sect. 2.4 would have been in place from the beginning of the pandemic, its global economic repercussions could have been less severe and much of the increase in global poverty could have been prevented. This is to say, they are now more needed than ever.

Changes in the IFFS are not acts of charity, but necessary to return to sustainable and equitable global growth, without which gains for some countries and country groupings that now propagate a 'me first' attitude in health issues and a return to protectionism will not be sustainable in the long-term leading to lower global growth and to a reversal in the upward trend in achieving the Sustainable Development Goals by 2030.

References

Afesorgbor, S. K., van Bergeijk, P. A. G., & Demena, B. A. (2021). COVID-19 and the threat to globalization: An optimistic note. In E. Papyrakis (Ed.), *Covid-19 and international development*. Springer.

Chandrasekhar, C. P. (2021). The challenge of LDC debt. *Economic and Political Weekly, 56*(3). https://www.networkideas.org/news-analysis/2021/02/the-challenge-of-ldc-debt/. Accessed on 21 Mar 2021.

Chowdhury, A., & Jomo, K. S. (2021). *IMF, World Bank must urgently help finance developing countries*. Blog. Available at: https://www.ksjomo.org/post/imf-world-bank-must-support-developing-countries-recovery. Accessed on 30 Mar 2021.

Development Initiatives. (2021). *Aid data 2019–2020: Analysis of trends before and during COVID-19*. Briefing. Available at: https://devinit.org/resources/aid-data-2019-2020-analysis-trends-before-during-covid/#downloads. Accessed on 20 Feb 2021.

FACTI Panel. (2021). *Financial Integrity for Sustainable Development*. Report of the High-Level Panel on International Financial Accountability, Transparency and Integrity for Achieving the 2030 Agenda. Financial Accountability, Transparency and Integrity (FACTI), New York. Available at: https://www.factipanel.org/explore-the-report. Accessed on 20 Mar 2021.

Gallagher, K., & Carlin F. M. (2020). The role of IMF in the fight against COVID-19: The IMF Covid Response Index. *Covid Economics* 42, 19 August 2020.

Gallagher, K., Gao, H., Kring, W., Ocampo, J. A., & Volz, U. (2021). Safety first: Expanding the global financial safety net in response to COVID-19. *Global Policy, 12*(1), 140–148.

Goolsbee, A., & Syverson, C. (2020). *Fear, lockdown and diversion: Comparing drivers of pandemic economic decline 2020* (NBER Working Paper 27432). National Bureau of Economic Research.

Haan, R. (1971). *Special drawing rights and development*. Stenfert Kroese.

Herman, B., Ocampo, J. A., & Spiegel, S. (2010). *Overcoming developing country debt crises*. Oxford University Press.

ICRIT. (2020). *The Global pandemic, sustainable economic recovery, and international taxation*. Independent Commission for the Reform of International Taxation. Available at: https://

www.icrict.com/icrict-documentsthe-global-pandemic-sustainable-economic-recovery-and-international-taxation. Accessed on 20 Mar 2021.
ILO. (2021). *ILO Monitor: COVID-19 and the world of work. Seventh edition updated estimates and analysis*. International Labour Organization, Geneva, Switzerland. Available at: https://www.ilo.org/global/topics/coronavirus/impacts-and-responses/WCMS_767028/lang%2D%2Den/index.htm. Accessed on 25 Jan 2021.
IMF. (2021a). *World Economic Outlook*, January 2021. International Monetary Fund. Avalable at: https://www.imf.org/en/Publications/WEO/Issues/2021/01/26/2021-world-economic-outlook-update. Accessed on 25 Apr 2021.
IMF. (2021b). *Fiscal monitor. Database of country fiscal measures in response to the COVID-19 pandemic*. International Monetary Fund. Data available at: https://www.imf.org/en/Topics/imf-and-covid19/Fiscal-Policies-Database-in-Response-to-COVID-19. Accessed on 09 Apr 2021
INET. (2021). *The pandemic and the economic crisis: A global agenda for urgent action* (Interim report of the commission for global economic transformation). Institute for New Economic Thinking. Available at: https://www.ineteconomics.org/research/research-papers/the-pandemic-and-the-economic-crisis-a-global-agenda-for-urgent-action. Accessed on 20 Mar 2021
Laborde, D., Martin, W., & Vos, R. (2020). *Impacts of COVID-19 on global poverty, food security and diets* (IFPRI Discussion Paper 01993 December). International Food Policy Research Institute.
Laborde, D., Martin, W., & Vos, R. (2021). Impacts of COVID-19 on global poverty, food security and diets. *Agricultural Economics, 52*(3), 375–390. Available at: https://doi.org/10.1111/agec.12624.
Mahler, D., Lakner, C, Castaneda Aguilar, R., & Wu, H. (2020). *The impact of COVID-19 (Coronavirus) on global poverty: Why Sub-Saharan Africa might be the region hardest hit?* World Bank Data Blog. Available at: https://blogs.worldbank.org/opendata/updated-estimates-impact-covid-19-global-poverty. Accessed on 20 Apr 2021.
Mukhtarov, M., Papyrakis, E., & Rieger, M. (2021). Covid-19 and water. In E. Papyrakis (Ed.), *Covid-19 and international development*. Springer.
Murshed, S. M. (2021). Consequences of the Covid-19 pandemic for economic inequality. In E. Papyrakis (Ed.), *Covid-19 and international development*. Springer.
Ocampo, J. A. (2015). Reforming the international monetary and financial architecture. In J. A. Alonso & J. A. Ocampo (Eds.), *Global governance and rules for the post-2015 era* (pp. 41–72). Bloomsbury Academic.
Oxfam. (2020). *Behind the numbers: A dataset on spending, accountability and recovery measures included in IMF Covid-19 loans. Online database*. Oxfam. Available at https://www.oxfam.org/en/international-financial-institutions/imf-covid-19-financing-and-fiscal-tracker. Accessed on 27 Mar 2021
Oxfam. (2021a). *The inequality virus* (Oxfam Briefing paper (January)). Oxfam. Avalable at https://oxfamilibrary.openrepository.com/bitstream/handle/10546/621149/bp-the-inequality-virus-250121-en.pdf. Accessed on 27 Mar 2021
Oxfam. (2021b). *Austerity is not a fair shot: 4 observations on the IMF's latest COVID 19 loans* (Oxfam Briefing paper (April)). Oxfam. Available at: https://medium.com/@OxfamIFIs/austerity-is-not-a-fair-shot-fdc4e11a06b6. Accessed on 30 Mar 2021
Sumner, A., Hoy, C., & Ortiz-Juarez, E. (2020). *Estimates of the impact of COVID-19 on global poverty* (WIDER Working Paper 2020/43). World Institute for Development Economics Research. Available at: https://www.wider.unu.edu/sites/default/files/Publications/Working-paper/PDF/wp2020-43.pdf. Accessed on 14 Apr 2021
UN-DESA. (2021). *World economic situation and prospects* (Monthly Briefing No 146 (5 February 2021)). United Nations Department of Economic and Social Affairs. Available at: https://www.un.org/development/desa/dpad/publication/world-economic-situation-and-prospects-february-2021-briefing-no-146/. Accessed on 03 Apr 2021

United Nations. (2012). *World Economic and Social Survey 2012: In search for new international development finance*. United Nations. Available at: http://www.un.org/en/development/desa/policy/wess/wess_current/2012wess.pdf. Accessed on 10 Apr 2021

United Nations. (2021). *Liquidity and debt solutions to invest in the SDGs: The time to act is now* (UN secretary-general policy brief (March 21)). United Nations. Available at: https://www.un.org/sites/un2.un.org/files/sg_policy_brief_on_liquidity_and_debt_solutions_march_2021.pdf. Accessed on 12 Mar2021

Van Bergeijk, P. (2021). *Pandemic economics*. Edward Elgar Publishers.

van der Hoeven, R. (2020). Multilateralism, employment and inequality in the context of COVID-19. In *UN committee for development policy (UNCDP), 2020, development policy and multilateralism after COVID-19*. United Nations.

Vos, R. (2017). From billions to trillions: Towards reform of development finance and the global reserve system. In P. van Bergeijk & R. van der Hoeven (Eds.), *Sustainable development goals and inequality*. Edward Elgar.

World Bank. (2020a). *Poverty and shared prosperity 2020: Reversals of fortune*. The World Bank. Available at: https://openknowledge.worldbank.org/handle/10986/34496. Accessed on 06 Mar 2021

World Bank. (2020b). *Global economic prospects 2020*. The World Bank. Available at: https://www.worldbank.org/en/publication/global-economic-prospects. Accessed on 03 Mar 2021

Chapter 3
COVID-19 and the Threat to Globalization: An Optimistic Note

Sylvanus Kwaku Afesorgbor, Peter A. G. van Bergeijk, and Binyam Afewerk Demena

Abstract We analyze the impact of COVID-19 on the world economic system through the three lenses of globalization, discussing economic, social and political aspects, respectively. The pandemic and the policy responses have hit these aspects to different degrees.

- Economically, the multilateral system has been under pressure; the quick recovery of world merchandise trade stands out, but FDI remains subdued.
- Socially, the reduction in tourism is the largest shock but here a sharp recovery is possible.
- Politically, the end of US membership of the WHO and the difficulty of global economic coordination in the G20 are key drivers. The election of a new US President allows for a quick reversal.

The outlook for openness of the world economy is still very much better than in the 1930s. Yes, deglobalization exists. Yes, overall globalization will probably be lower for the foreseeable future. Our societies will, however, remain much more open than at the start of the globalization wave in 1990, connected via the internet with an intensity never seen before in history.

S. K. Afesorgbor
Department of Food, Agricultural and Resource Economics, University of Guelph, Guelph, ON, Canada

P. A. G. van Bergeijk (✉) · B. A. Demena
International Institute of Social Studies, Department of Development Economics, Erasmus University Rotterdam, The Hague, The Netherlands
e-mail: bergeijk@iss.nl

© The Author(s), under exclusive license to Springer Nature Switzerland AG 2022
E. Papyrakis (ed.), *COVID-19 and International Development*,
https://doi.org/10.1007/978-3-030-82339-9_3

3.1 Introduction

Globalization has been referred to as a multifaceted concept that describes the process of creating networks of connections among actors at intra- or multicontinental distances (Gygil et al., 2019). According to these authors, it is a more holistic concept that captures the increased intercedence of national economies, and the trend towards greater integration of different varieties of flows such as information, goods, labour and capital markets. Because of this broad definition of globalization, continuous attempts have been made starting with Dreher (2006) to develop a more comprehensive composite index such as the KOF Globalization Index that encompasses three main dimensions of the process of globalization, namely, economic, political and social dimensions. Increased globalization has been a constant feature of the world economy since the 1970s and especially since 1990 after the fall of the Iron Curtain and with China gathering speed.

More recently, however, there has been growing discontent and negative sentiments about the effect of globalization (Stiglitz, 2002, 2018). These negative sentiments have manifested in different ways through the election of President Donald Trump, the Brexit and attacks on the World Trade Organization (WTO). Under the aegis of then-President D. Trump who not only staged an attack on the WTO by refusing to appoint new members to the WTO's Appellate Body, but also launched trade wars against China, North American Free Trade Agreement (NAFTA) and the European Union (EU), putting further strains on world trade and international cooperation. Afesorgbor and Beaulieu (2021) argue the Trump presidency strained diplomatic relationship with close allies and undermined the rules-based global system, and this therefore created uncertainty for global trade.

These occurrences that constitute a major setback to the pace of globalization set the stage of growing protectionism and nationalism in the world. These actors were political, but more recently the outbreak of the COVID-19 pandemic introduces new – medical – threat to globalization emanating from health risk posed by the contagious nature of the COVID-19. In a sense the pandemic reflects both globalization (a virus went global in a few weeks thanks to high level of globalization and interconnectedness (Lipscy, 2020) and deglobalization (the breakdown of international cooperation and the re-emergence of zero-sum thinking and raw beggar-thy-neighbour polices on the markets for medical productive gear, medical machinery and vaccines).

The COVID-19 outbreak provided a threat to both lives and livelihood (UNCTAD, 2020) which are the basic foundation for national economies to thrive and thereby triggered an effect for the global economy. Because of the strong and interdependent global production linkages coupled with the closure of international borders, businesses, and factories, the economic expectations and forecasts were generally pessimistic. The prospect of the world plunging into another major and long-term economic recession comparable to the Great Depression in 1929/30 and the Great Recession of 2007/8 was on the mind of many economists and international organizations.

3 COVID-19 and the Threat to Globalization: An Optimistic Note

Indeed, in April 2020, the outlook for the world economy and especially the world trading system was dismal. The outbreak of the COVID-19 pandemic came on top of that and led to an unprecedented dive of world trade by 15% in April 2020 and the trade collapse initially was stronger than previous episodes of downfall. During the first wave of the Covid-19 pandemic projections for the contraction of the annual volume of world trade in 2020 ranged from minus 10 to minus 32% (see Fig. 3.1). The outlook for globalization was grim. The pandemic at the same time highlighted the importance of global interdependence and the awareness of the need to develop a collective global strategy (Committee for Development Policy-CDP, 2020). Initially, the occurrence of the viral infection was thought to be only a problem for China but within a couple of weeks the outbreak of the novel COVID-19 virus became first an international problem and soon afterwards a global issue and was officially announced a pandemic. By symmetry, the eradication of the viral spread will require international cooperation and a global effort such that no single country is left behind to form a pool for new variants and future outbreaks. Vaccines must be made available and affordable to all countries. Just as globalization has ramifications for all countries, the health of different nations is intertwined as a pandemic is a 'public bad': the health of one nation affects the health of the other (van Bergeijk, 2013; CDP, 2020).

The reports of the expected death of globalization, however, were – with hindsight – grossly exaggerated. Admittedly, the fall in world trade had initially been exceptionally strong indeed. Recovery, however, set in early compared to the two major historical episodes of trade collapse during the Great Recession and the Great Depression suggesting stronger resilience of world trade than anticipated by the international organizations. Figure 3.2 compares the developments of the world merchandise trade volume during the three major world trade collapses that occurred during the Great Depression in the 1930s, the Global Financial crisis in 2008/9 and the lockdowns during the first wave of the COVID 19 pandemic.

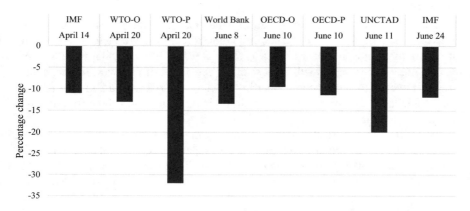

Fig. 3.1 Forecasts for real world trade growth in 2020. (Note: O Optimistic scenario, P Pessimistic Scenario)

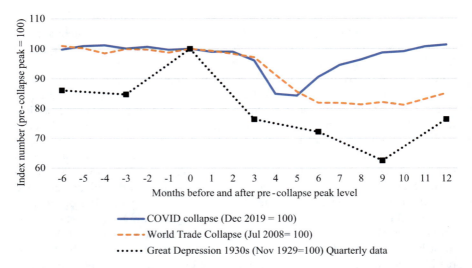

Fig. 3.2 Comparing the COVID collapse to the 2008/9 world trade collapse and the Great Depression. (*Sources:* Updated from Meijerink et al., 2020)

The comparison starts half a year before the peak level of trade just before the collapse that we use to identify period 'zero'. The contractions in 2008/9 and 2020 both started to bite after a quarter but in the fourth month of the COVID collapse world trade was already 15% below previous peak level, while this was 'only' 8% during the 2008/9 World Trade Collapse. Global merchandise trade in May 2020, however, bottomed out, and full recovery was achieved in November 2020. At the comparable point of the 2008/9 trade collapse the volume of world trade was still 17% below pre-crisis peak level. In the 1930s world trade one year into the collapse was 28% below pre-crisis peak level. Clearly, the world trade collapses before the COVID-19 contraction were both deeper and much more protracted (van Bergeijk, 2019, 2021).

The resilience of world trade during the 'natural experiment' of the Great Recession is remarkable and the starting point of research into factors determining its elasticity (van Bergeijk et al., 2017). The COVID-19 trade collapse adds a second 'natural experiment' with even stronger resilience of real-world merchandise trade. However, as will become clear in this chapter, 2020 was a bad year for globalization and therefore we need to be careful in our conclusion about the pandemic threat for globalization and the future outlook.

We will first discuss the developments during the COVID-19 pandemic. We organize this discussion of developments into separate sections, that recognize that deglobalization, like globalization, has several dimensions (Dreher et al., 2008; Gygli et al., 2019).[1] This is important as the threat of COVID-19 would have

---

[1] We deviate from the common procedure in the literature on globalization to include remittances and (official development) assistance

heterogeneous effect on different dimensions of globalization. In addition, the effect would also differ for different regions or countries depending on their level of exposure to the viral infection and the integration of the region or country in the world economy. Apart from that, the COVID-19 effect would also have differential effect on different industries, sectors or products. For instance, UNCTAD (2020) shows that there was upsurge (as would be expected) in global merchandise trade of medical production in the second quarter of 2020 during the peak of the pandemic. Clearly, in analyzing the effect of COVID-19 on globalization, one must take different dimensions, products and countries into perspective. UNCTAD (2020) further highlights the heterogenous effect of COVID-19 crisis on different dimensions of economic globalization such as trade, FDI, global production and employment as well as the effect for different countries.

This chapter is organized as follows. Section 3.2 discusses the economic dimension that relates to interaction between the private sectors in different countries (or within multinational firms) and covers such items as foreign trade (exports and imports of goods and services), private capital flows (including bank lending, portfolio investment, Foreign Direct Investment - FDI). The social dimension is discussed in Sect. 3.3 and covers interactions between (groups of) individuals from different countries as being shaped by tourism, migration, remittances as well as cultural and personal exchanges. Section 3.4 focusses on the political dimension of globalization, that is the interactions between States (both bilaterally and multilaterally) as observed by (changes in) their membership of international institutions, involvement in Treaties, participation in peace-keeping missions and development assistance. Having sketched the developments during 2020 we analyze the post-COVID 19 outlook in Sect. 3.5.

3.2 Economic Globalization and COVID-19

Economic globalization has been conceptualized by means of flows of goods, services, capital and information in connection to long distance market transactions (Dreher, 2006). For the economic dimension of globalization, Gygli et al. (2019) further distinguish between trade and financial globalization. They measure economic globalization using *de facto* variables such as foreign trade (exports and imports of goods and services) and *de jure* factors including trade regulations, tariffs and trade agreements. Similarly, private capital flows (including bank lending, portfolio investment, FDI) were used as *de facto* measures of financial globalization while the restriction on capital flows, capital account openness and international investment are de jure measures of financial globalization.

How the outbreak of COVID-19 would affect these different indicators under economic globalization can differ. Already, van Bergeijk (2019, 2021) argued that prior to the COVID-19, many developed countries were experiencing lower levels of economic globalization as a result of the anti-globalization wave that has engulfed Global North. The pandemic would likely deepen the impetus of the deglobalization trend and could even have future implications. The onset of the COVID-19 crisis created panic-buying in many countries resulting in acute shortages of products on shelves of most retail companies. Such occurrences create the notion of over-dependence of the domestic supply chain on foreign suppliers. This could, consequently, be used as a subtle call to enact protectionist policies in order to build the capacity of domestic industries to promote self-sufficiency in the production of agricultural and manufacturing goods (Kerr, 2020).

Although the pandemic is global, different regions and countries experienced differential effect on various indicators of the economic dimension of globalization. In particular, different regions or countries are exposed differently as they are integrated into the global economy at different intensities. The level of integration determines the size of the impact of the pandemic on international trade, capital flows and FDI. Therefore, regions or countries that are well integrated in the global economy experience higher rate of infections. This seems to be the case as many less developed countries (LDCs) especially African countries that are less integrated in the world economy have significantly low rate of infections.

3.2.1 Merchandise Trade

Figure 3.3 shows the heterogeneity of merchandise exports and imports before COVID-19 outbreak and during the pandemic. Although the figure shows a general decline in growth rates for merchandize trade for all countries, the rate of decline was most pronounced for the advanced economies as detailed in Table 3.1.

3.2.2 Trade in Services

Figure 3.3 and Table 3.1 deal with merchandise trade only and miss an important part of the puzzle: international trade in services, as it accounts for 25% of total trade (Shingal, 2020). The COVID-19 pandemic could be expected to be less deleterious for trade in services based on the argument that service trade is less sensitive to demand shocks and less dependent on supply finance (Ariu, 2016). However, Shingal (2020) argues that social-distancing and contagion-related fears would hinder trade in services, as services transactions such as health and education services, movement of workers (IT professionals requiring to work onsite) and especially tourism require some form of physical proximity between buyers and sellers. Preliminary statistics from UNCTAD (2020) also shows a greater negative effect for services trade. The statistics show that globally services trade contracted by 21% in the second quarter of 2020 compared to 18% decline in trade in goods within the same period.

3 COVID-19 and the Threat to Globalization: An Optimistic Note

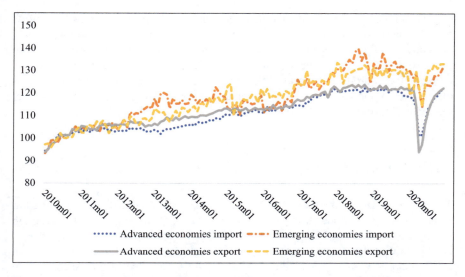

Fig. 3.3 Global and regional patterns of merchandise trade. (*Source:* CPB World Trade Monitor, accessed March 2021)

Table 3.1 Monthly changes of the volume of international merchandise trade (2020)

	Import		Export	
	Advanced economies	Emerging economies	Advanced economies	Emerging economies
January	−0.6%	−0.2%	−0.9%	−7.5%
February	−0.2%	−2.2%	1.4%	0.3%
March	−3.4%	1.3%	−6.2%	7.6%
April	−11.3%	−5.4%	−18.3%	−5.5%
May	0.7%	−6.4%	3.6%	−6.6%
June	5.7%	8.7%	9.4%	8.6%
July	5.3%	−0.3%	5.4%	4.9%
August	2.1%	−0.2%	3.4%	−0.2%
September	2.1%	3.6%	2.4%	2.3%
October	0.7%	−0.2%	1.1%	−1.5%
November	2.3%	1.4%	1.1%	1.5%
December	−0.6%	2.1%	0.7%	0.1%

Source: CPB World Trade Monitor, accessed March 2021

3.2.3 Trade by Product

Breaking down the effect of decline in merchandise exports across the different products or sectors shows that the effect of pandemic was heterogenous. In particular, sectors such as automotive and chemical sectors declined significantly in major trading economies like the US, Europe and China while sectors within the textile, office machinery and precision instruments actually witnessed an increase in exports from China (UNCTAD, 2020). Trade in medical related products including personal protective equipment, sanitizers and ventilators witnessed a particularly sharp increase of 186% in the second quarter of 2020 and there was similar growth in home office equipment such as Wi-Fi routers, laptops and portable storage. It is often said that globalization has winners and losers; clearly the same can be said at the sectoral level for the COVID-19 pandemic.

3.2.4 Foreign Direct Investment (FDI)

The heterogeneity of the COVID-19 impact on international flows is not limited to merchandise and services trade and also occurs for FDI flows (Fig. 3.4). Although the trajectory of falling FDI predates the COVID-19 pandemic reflecting that deglobalization was already predating the pandemic, the rapid and significant decline in 2020 stands out and exceed the decline of the Great Recession. According to UNCTAD (2021), the impact of COVID-19 on FDI was immediate, as global FDI flows declined almost by nearly half in 2020. Figure 3.4 shows that COVID-19 bit into FDI inflows by 42% globally (see also Fig. 3.5). This can be compared to the global financial crisis that started in September 2007, but global FDI contracted largely later in 2009 by 39%, plummeted at an accelerated rate lately (Demena, 2017). Figure 3.4 also shows the immediacy of the COVID-19 impact on the global FDI. Clearly, the impact of the COVID-19 FDI contraction is both deeper and more immediate as compared to the Global Financial Crisis. Again, this decline was mostly concentrated in developed countries, with only marginal decline for the group of developing countries. As forecasted by UNCTAD (2021), the latter is less than expected.

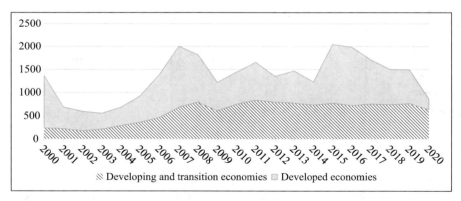

Fig. 3.4 FDI inflows: global and by group of economies, 2007–2020 (billions of US dollars) *Source*: Demena (2017) and UNCTAD (2021). (Note: All FDI figures are billions of US dollars at current prices and current exchange rates)

3 COVID-19 and the Threat to Globalization: An Optimistic Note   37

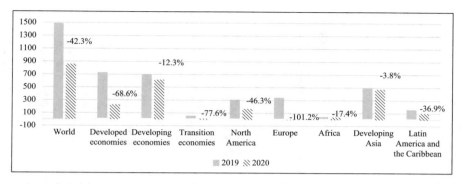

Fig. 3.5 Foreign direct investment inflows across different regions. (*Source*: UNCTAD, 2021. Note: See Fig. 3.4)

Figure 3.5 shows the decline in FDI across different regions of the world. Developed nations experienced 69% decrease in 2020 compared to 12% decline for developing economies. Unexpectedly, the impact among the transition economies were very sharp (about 77%), and UNCTAD suggested that this is mainly due to the decline of the flows to the Russian Federation. Looking ahead, according to the forecast of the UNTAD (2021), the pattern of FDI is expected to remain weak to recover in 2021, indicating the outlook remains dire. Thus COVID-19 impact will continue to provide a downward pressure on FDI flows.

3.3 Social Globalization and COVID-19

There are three different sub-dimensions of social globalization and these include interpersonal, informational and cultural aspects of globalization (Dreher, 2006). According to Gygli et al. (2019) interpersonal globalization involves personal links and/or interaction with foreign nationals through events such as migration, international telephone calls and international remittances paid or received by citizens. Informational globalization measures the actual flow of ideas, knowledge and images and it is measured using key variables such as internet bandwidth, international patents and technology exports. The cultural aspect of social globalization also relates the penetration of foreign cultural products such as franchise acquisition or foreign trademarks applications. Keohane and Nye (2000) argue that social globalization is the most pervasive aspect of globalization. Relating the COVID-19 to social globalization is thus important and also because the sub-dimensions could be affected differently. First, COVID-19 reduces interpersonal globalization. Many countries impose travel restriction of both residents and foreign nationals moving in and out of the country. The border closure would directly hinder the temporary migration, especially tourist and foreign students' movement in and out of countries.

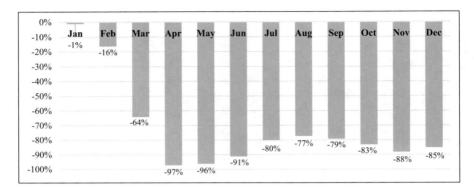

Fig. 3.6 International (World) tourist arrivals (% change monthly comparison between 2020 and 2019). (*Source*: UNWTO: https://www.unwto.org/)

Figure 3.6 confirms the possible negative effect on international tourist arrivals in the world. Comparing the percentage growth month-over-month for 2019/2020, Fig. 3.6 shows 96–97% decrease during the peak of the COVID pandemic and lockdowns. Although this has reduced marginally over the years the percentage decline remains substantially large even by December 2020 (85% lower than as compared to December 2019). It were not only the travel restrictions imposed by many governments that dissuaded tourist travels; most people may have also voluntarily decided to cease travels for personal safety reasons because even when the travel restrictions or border closures were removed as suggested by the reported developments in the fourth quarter of 2020: there was still limited number of tourist travels in 2020 compared to 2019. According to the World Tourism Organization (UNWTO) world tourism barometer,[2] at the beginning of 2021 the demand for international travel remained very weak (87% below the January 2019). At the time of writing, amid new outbreaks and tighter travel restrictions along with the slower and quite uneven distribution of vaccination roll-out across countries and regions a continued slowdown was expected. The UNWTO outlined two scenarios for recovery in 2021 as compared to the historic lows of 2020. In the optimistic scenario international travel would rebound in July 2021, leading to a partial recovery by 66%. In the pessimistic scenario, the recovery sets in September 2021, resulting to a 22% recovery. On balance international arrivals remain 50–70% below the 2019 level of international arrivals.

Migrant remittances were also affected not because of any formal restrictions on remittances but mainly because of a negative labour market shock on immigrant employment. The latter is also confirmed by Chap. 6 on the short-term impact of COVID-19 on labour market outcomes of this book (Demena et al., 2021); the authors find that, overall, the pandemic negatively impacts various labour market outcomes, especially in developed economies, which subsequently reduces the amount of remittances that can become repatriated to developing economies.

---

[2] https://www.unwto.org/unwto-world-tourism-barometer-data accessed 1 April 2021.

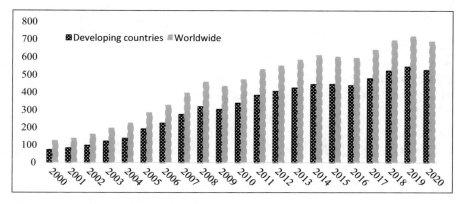

Fig. 3.7 Migrant remittance inflows (US$ Billion). (*Source*: KNOMAD – https://www.knomad.org/)

Orozco (2002) argues that migration represents an important dimension of globalization and family remittances constitute a major factor in integrating. Borjas and Cassidy (2020) find that in the US labour market, COVID-19 related labour market disruptions severely affected immigrants. They explain that joblessness was higher among immigrants because they were less likely to work in jobs that could be performed remotely. This stylized fact was also confirmed by Capps et al. (2020), who explained that COVID-19 severely affected industries that had higher concentration of immigrants especially Latinos in the US. Figure 3.7 shows the long-term increase in remittances inflow worldwide and also for developing countries, as well as the sharp decrease during COVID-19 era. There was a similar decline for remittances following the global financial crisis in 2008/9. It is unclear when recovery sets in although recovery after vaccinations could be quick.

3.4 Political Globalization and COVID-19

Political globalization captures the ability of countries to engage in international political cooperation as well as the diffusion of government policies (Gygli et al., 2019). Key variables used in measuring political globalization include membership of international institutions, involvement in treaties, participation in peace-keeping missions and development assistance. These various indicators measure how much a government is influenced by foreign counterparts and resources.

The initial outbreak of COVID-19 pandemic affected international cooperation negatively mainly because of the blame game between the two largest economies in the world, the US and China. The Trump administration blamed China for being less transparent, failing to contain COVID-19 virus and allowing it to spread to other countries. This was also followed by counter-accusations from Chinese officials, who also blamed the US for mismanaging and failing to take the pandemic seriously (Horsley, 2020). In addition, the Trump presidency undermined global cooperation and global public health when the US cut financial support to the World Health Organization (WHO) during the middle of the pandemic. Brown and Susskind (2020) viewed the rising anti-globalization sentiments and nationalism (Make America Great Again, Brexit) as impediments that undermined the willingness of countries to coordinate their responses in the fight against the pandemic.

3.4.1 Global Public Good

The occurrence of COVID-19 also presents an opportunity to correct earlier skirmishes and for a more global cooperation under the aegis of the WHO in combating the virus. Brown and Susskind (2020) have argued that the control of the COVID-19 pandemic should be treated as a global public good. The increasing level of globalization and interdependence of nations make it imperative for the world to treat COVID-19 prevention as a global public good; if a single country fails to control the virus within its borders, the infection can easily spread beyond its borders to the rest of the world (Brown & Susskind, 2020). Additionally, international cooperation in vaccine development and distribution are equally important.

Although global cooperation did not start immediately with the outbreak of COVID-19, there were many efforts later by different countries to cooperate in fighting the pandemic. European countries provided medical gear to China in the early phase of the pandemic. Later on, China supported adversely affected countries like Italy which became the epicentre of the COVID-19 pandemic in Europe. The medical assistance and supplies aid from China to Italy amounted to about 31 tons of medical supplies during the peak of the crisis.[3] Medical assistance from China was received with mixed reactions mainly because it was also believed that the gesture was divisive for the Europe and such acts would play to Chinese geopolitical advantage within the region.[4] On the medical front, Chinese scientists played a significant role in the sequencing of the genome and development of tests that were shared with other health facilities across the world (and, hence, helped in detecting infection cases and manage the spread of the virus; Horsley, 2020).

Similarly, international assistance of medical aid from countries like Russia sent to Italy and even the US (Biscop, 2020) signals the importance of global partnership in eradicating the pandemic. Similar medical aid was also transferred from Cuba to support medical personnel in Italy. These gestures of international medical assistance also point to a concerted global effort in curbing the spread of the COVID-19 virus. Politically, the outbreak of the corona virus could be used as a building block to reinforce international cooperation in the world and strengthen the pillars of

---

[3] https://foreignpolicy.com/2020/03/14/coronavirus-eu-abandoning-italy-china-aid/, accessed 31 March 2021.

[4] https://foreignpolicy.com/2020/03/14/coronavirus-eu-abandoning-italy-china-aid/, accessed 31 March 2021.

political globalization. Brown and Susskind (2020) cite the EU-led global health pledging conference that was committed to raise $8 billion to fund the activities of global institutions such as the Global Alliance for Vaccines and Immunization (GAVI) and the WHO that are responsible for coordinating the global response to the virus. In coordinating their efforts, many leaders of the G20 group of countries made commitments, including China who made a pledge to contribute $20 million to the fund.[5] In March 2021 the head of the WHO together with a group of world leaders initiated a discussion calling for international treaty aimed at enhancing pandemic preparedness and pandemic response with a view to new diseases in the post COVID-19 world.

3.4.2 Development Cooperation

The other sphere of political globalization is development related foreign aid or assistance. Given that the COVID-19 pandemic negatively impacted other sources of development finance as presented above, such as FDI, remittances and trade, countries in Global South were likely in need of increasing foreign aid.

According to the OECD,[6] external financing flows to developing economies fell by US$ 700 billion in 2020 as compared to 2019, representing a drop by 45%. The report further argued that this reduction in external financing greatly exceeds the immediate impact of the global financial crisis by about 60%. Another recent study by Adam et al. (2020) indicated that aid in development assistance needs to double in order for sub-Saharan Africa to recover. Responses to the substantial gap in development finance triggered by the COVID-19 impact appear to be promising. The World Bank has pledged to make $160 billion available in grants and financial support over a 15-month period to help 100 developing countries respond to the health, social and economic impacts of COVID-19.[7] Apart from this, individual countries are also extending economic and development assistance to poor countries. For instance, the Netherlands expanded its development financing to developing countries by US$ 354 million in September 2020 and plans to contribute a further US$ 548 million for the upcoming years.[8] On March 2021, the US sets to lead the world in filling the financial aid gap by allocating US$11 billion in International aid in the latest COVID Bill.

---

[5] https://healthpolicy-watch.news/countries-pledge-7-36-billion-euro-towards-global-covid-19-response/, accessed 31 March 2021.
[6] The impact of the coronavirus (COVID-19) crisis on development finance," Paris: Organisation for Economic Co-operation and Development, 24 June 2020, http://www.oecd.org/coronavirus/policy-responses/the-impact-of-the-coronavirus-covid-19-crisis-on-development-finance-9de00b3b/ accessed 1 April 2021.
[7] https://www.worldbank.org/en/news/press-release/2020/05/19/world-bank-group-100-countries--get-support-in-response-to-covid-19-coronavirus, accessed 2 April 2021.
[8] https://borgenproject.org/the-netherlands-foreign-aid/ accessed on 1 April 2021.

Brown (2021:2) has argued that the COVID-19 pandemic has greatly accelerated significant positive trends, weakening the perception that foreign aid essentially flows from the Global North to the Global South and reinforcing awareness of the importance of joint efforts for global public goods and humanitarian assistance. He further argues that there is renewed emphasis on well-being, instead of aid favouring investment in growth (with the latter having more uncertain benefits for the global poor). There is also a need to consider the international fiscal and financial system. Van der Hoeven and Vos (2021) in Chap. 2 argue that the global consequences of COVID-19 would have been less severe and that much of the increase in global poverty could have been prevented and propose a comprehensive package of international tax reform, sovereign debt restructuring and relief, a reform of policy conditionality and an increase in the IMF's special drawing rights geared towards developing countries and emerging markets.

3.5 Outlook for Globalization After COVID-19: A Sense of Optimism

The first reason for optimism is the noteworthy resilience of world merchandise trade and investment during previous global crises. Multinational enterprises have already had their stress test during the 2008/9 collapse of world trade. That collapse kickstarted the process of deglobalization, but as illustrated in Fig. 3.2 global merchandise trade and industrial production recovered to previous peak quickly and even quicker during the COVID-19 crisis. This is the big difference with the Great Depression of the 1930s and it may be related to the fact that world trade is governed and supported by the multilateral trading system. The shock was sharp and immediate, but so was the recovery. The so-called immaterial flows (FDI, remittances, tourism, official development cooperation) have been hit harder and recovery is not to be expected before vaccination is sufficiently 'truly global', but the expectation of a speedy recovery is realistic at the time of writing.

The attacks on supranational governance and international cooperation were a symptom of an underlying disease that paradoxically may have been cured by the pandemic. The deglobalization process was driven by greater inequality and a lacklustre trickling down of the benefits of international trade and investment. We have learned that inequalities are the breeding ground for the spreading of disease and the suffering that may follow. Reducing epidemic vulnerabilities requires reducing such inequalities. But fighting potential next pandemics implies that we cannot limit our attention to inequalities at home, because the equalities around the world – within and between countries – provide breeding grounds and disease pools from which new variants, viruses and other contagious diseases emerge. The implication is that reducing inequalities in other countries and continents becomes a business proposition: an investment project with a high rate of return.

And last but not least, the outlook for openness of the world economy is still very much better than in the 1930s. Yes, deglobalization exists. Yes, overall globalization will probably be lower for the foreseeable future. Our societies will, however, remain much more open than at the start of the globalization wave in 1990, connected via the internet with an intensity never seen before in history.

References

Adam, C., Henstridge, M., & Lee, S. (2020). After the lockdown: Macroeconomic adjustment to the COVID-19 pandemic in sub-Saharan Africa. *Oxford Review of Economic Policy, 36*, S338–S358.
Afesorgbor, S. K., & Beaulieu, E. (2021). Role of international politics on Agri-food trade: Evidence from US–Canada bilateral relations. *Canadian Journal of Agricultural Economics/Revue canadienne d'agroeconomie, 69*(1), 27–35.
Ariu, A. (2016). Crisis-proof services: Why trade in services did not suffer during the 2008–2009 collapse. *Journal of International Economics, 98*, 138–149.
Biscop, S. (2020). *Coronavirus and power: The impact on international politics*. Egmont Institute.
Borjas, G. J., & Cassidy, H. (2020). *The adverse effect of the covid-19 labour market shock on immigrant employment (No. w27243)*. National Bureau of Economic Research.
Brown, S. (2021). The impact of COVID-19 on development assistance. *International Journal*, 0020702020986888.
Brown, G., & Susskind, D. (2020). International cooperation during the COVID-19 pandemic. *Oxford Review of Economic Policy, 36*, S64–S76.
Capps, R., Batalova, J., & Gelatt, J. (2020). *COVID-19 and unemployment: Assessing the early fallout for immigrants and other US workers*. Washington, DC: Migration Policy Institute.
Committee for Development Policy. (2020). *Development policy and multilateralism after COVID-19*. United Nations.
Demena, B. A. (2017). *Essays on intra-industry spillovers from FDI in developing countries: A firm- level analysis with a focus on Sub-Saharan Africa*, PhD diss., Erasmus University, The Hague.
Demena, B. A., Floridi, A., & Wagner, N. (2021). The short-term impact of COVID-19 on labour market outcomes: Comparative systematic evidence. In E. Papyrakis (Ed.), *Covid-19 and international development*. Springer.
Dreher, A. (2006). Does globalization affect growth? Evidence from a new index of globalization. *Applied Economics, 38*(10), 1091–1110.
Dreher, A., Gaston, N., & Martens, P. (2008). *Measuring globalisation. Gauging its consequences*. Springer.
Gygli, S., Haelg, F., Potrafke, N., & Sturm, J. E. (2019). The KOF globalisation index–revisited. *The Review of International Organizations, 14*(3), 543–574.
Horsley, J. P. (2020). *Let's end the COVID-19 blame game: Reconsidering China's role in the pandemic*. The Brookings Institution. Accessed 31st Mar 2021 from https://www.brookings.edu/blog/order-from-chaos/2020/08/19/lets-end-the-covid-19-blame-game-reconsidering-chinas-role-in-the-pandemic/
Keohane, R. O., & Nye, J. S., Jr. (2000). Globalization: What's new? What's not? (And so what?). *Foreign Policy, 118*, 104–119.
Kerr, W. A. (2020). The COVID-19 pandemic and agriculture: Short-and long-run implications for international trade relations. *Canadian Journal of Agricultural Economics/Revue canadienne d'agroeconomie, 68*(2), 225–229.
KOF Globalisation Index – Revisited', KOF Working Paper 439, ETH Zurich.
Lipscy, P. Y. (2020). COVID-19 and the politics of crisis. *International Organization, 74*(S1), 1–30.

Meijerink, G., Hendriks, B., & van Bergeijk, P. A. G., (2020). *Covid-19 and world merchandise trade: Unexpected resilience.* https://voxeu.org/article/covid-19-and-world-merchandise-trade

OECD. (2020). *The impact of the coronavirus (COVID-19) crisis on development finance.* OECD Publishing. http://www.oecd.org/coronavirus/policy-responses/the-impact-of-the-coronavirus-covid-19-crisis-on-development-finance-9de00b3b/

Orozco, M. (2002). Globalization and migration: The impact of family remittances in Latin America. *Latin American Politics and Society, 44*(2), 41–66.

Shingal, A. (2020). *Services trade and COVID-19. VOX EU.* Centre for Economic Policy Research (CEPR) Policy Brief. Accessed 28th Mar 2021 https://voxeu.org/article/services-trade-and-covid-19

Stiglitz, J. E. (2002). *Globalization and its discontents.* Norton.

Stiglitz, J. E. (2018). *Globalization and its discontents revisited. Anti-globalization in the era of trump.* Norton.

United Nation Conference on Trade and Development (UNCTAD). (2020). The *impact of COVID-19 on trade and development. Transitioning to a new normal.* Accessed 18 Mar 2021. https://unctad.org/system/files/official-document/osg2020d1_en.pdf

United Nation Conference on Trade and Development (UNCTAD). (2021). Investment Trends Monitor', Issue 38. Accessed 18 March 2021. https://unctad.org/system/files/official-document/diaeiainf2021d1_en.pdf

Van Bergeijk, P. A. G. (2013). *Earth economics: An introduction to demand management, long-run growth and global economic governance.* Edward Elgar.

van Bergeijk, P. A. G. (2019). *Deglobalization 2.0: Trade and openness during the great depression and the great recession.* Edward Elgar.

van Bergeijk, P. A. G. (2021). *Pandemic economics.* Edward Elgar.

van Bergeijk, P. A. G., Brakman, S., & van Marrewijk, C. (2017). Heterogeneous economic resilience and the great recession's world trade collapse. *Papers in Regional Science, 96*(1), 3–12.

van der Hoeven, R., & Vos, R. (2021). Reforming the international financial and fiscal system for better COVID-19 and post-pandemic crisis responsiveness. In E. Papyrakis (Ed.), *Covid-19 and international development.* Springer.

Chapter 4
Experiences of Eritrean and Ethiopian Migrants During COVID-19 in the Netherlands

Bezawit Fantu, Genet Haile, Yordanos Lassooy Tekle, Sreerekha Sathi, Binyam Afewerk Demena, and Zemzem Shigute

Abstract This chapter aims to shed light on the experiences of Eritrean and Ethiopian migrants during the COVID-19 pandemic in the Hague, the Netherlands. These include health (both physical and mental), economic and social effects. Experiences that were particular to women and children were also explored. Eighteen individual migrants varying in terms of their gender, country of origin (Eritrea or Ethiopia), profession, years of stay in the Netherlands, and marital status were interviewed using an in-depth interview guide. In addition, key informant interviews were held with representatives of two organizations working with migrant communities. An intersectionality lens was applied to frame the complex and interconnected challenges faced by migrants. Specifically, the concepts of precarious work and gender-based division of labor were used to frame findings related to financial impact and women's experience with the pandemic, respectively. Research findings revealed intersecting layers of struggle that pose challenges to the

---

B. Fantu
Girl Effect, Addis Ababa, Ethiopia

G. Haile
Dutch Refugee Council/Vluchtelingenwerk Nederland, Utrecht, The Netherlands

Y. L. Tekle
Cultuur in Harmonie, Zeewolde, The Netherlands

S. Sathi · B. A. Demena
International Institute of Social Studies, Erasmus University Rotterdam, Rotterdam, The Netherlands

Z. Shigute (✉)
International Institute of Social Studies, Erasmus University Rotterdam, Rotterdam, The Netherlands

Institute of Development and Policy Research, Addis Ababa University, Addis Ababa, Ethiopia
e-mail: shuka@iss.nl

© The Author(s), under exclusive license to Springer Nature Switzerland AG 2022
E. Papyrakis (ed.), *COVID-19 and International Development*,
https://doi.org/10.1007/978-3-030-82339-9_4

lives of these migrants based on various factors such as language skills, employment, gender, duration of stay in the Netherlands and marital status.

4.1 Introduction

Since the outbreak of COVID-19, countries around the world have unanimously embraced Non-Pharmaceutical Intervention (NPI) measures, including, maintaining physical distance, repeated handwashing, use of sanitizers and facemasks as preventive measures. Although these measures are encouraged by health professionals to maintain a functional health system, they have restrictive implications for an individual's way of life, social interaction, mental wellbeing, and the overall economic system (Vieira et al., 2020: 38).

It is perhaps self-evident, that the experience of various groups of people during the pandemic will differ based on their social, economic, cultural, and physical conditions. A particularly vulnerable group are migrants and potentially the pandemic and its associated restrictions are likely to exacerbate difficulties faced by them in host countries. As it is, migrants and refugees are consistently faced with difficulties such as insecurity, unemployment, and "the ramifications that come with the postponement of decisions on their legal status or reduction of employment, legal, and administrative services" (Kluge et al., 2020: 1238). Additionally, new and ever-changing information about prevention, risks, and treatment of COVID-19 are not easily available to migrants, who may hardly speak the local language and have not yet fully acculturated into their host countries (Kluge et al., 2020: 1238, see also Murshed, 2021).

It is important to shed light on these different experiences and to examine how disadvantaged groups, such as migrants, are potentially exposed to further layers of difficulties due to calamities such as coronavirus. Accordingly, this chapter investigates the experiences of Eritrean and Ethiopian migrants in the Netherlands based on in-depth interviews with eighteen individuals, from the two communities and representative of two organizations, namely Helpdesk Nieuwkomers and Pharos, which work with migrant communities. The eighteen interviewees were selected using snowball sampling method and interviewed from July to August 2020. Helpdesk Nieuwkomers is an initiative established at the start of the pandemic by four migrant-led organizations advocating for inclusion and empowerment of migrants. It is staffed with close to 40 migrant volunteers who answer daily hotline calls from newcomers speaking the two most common (migrant) languages, Tigrinya and Arabic. The volunteers provide information on various topics related to the corona pandemic. Pharos is an organization that works towards "reducing existing health disparities between different groups of people" (Pharos, 2020).

The chapter is organized in the following manner. After a brief overview on Eritrean and Ethiopian migrants in the Netherlands, findings on the intersecting challenges of Eritrean and Ethiopian migrants are provided in section three. Section four concludes the chapter with some policy remarks.

4.2 Eritrean and Ethiopian Migrants in the Netherlands: An Overview

Before the 1970s, migration of Ethiopians to the Netherlands was limited (Van Heelsum in Ong'ayo 2010: 78). In the 1980s, the number of Ethiopian migrants increased steadily due to drought and economic stagnation in the home country. The most recent and main reasons for Ethiopians to migrate to the Netherlands are political resistance and disputes with nations bordering Ethiopia (Ong'ayo, 2010: 78). In Eritrea, the youth appear to have adopted a 'culture of migration' where they adamantly believe migration is the only way to make a decent living for two major reasons - to evade recruitment in the country's military service and to make an economically better living. Since the economic conditions within Eritrea do not offer opportunities to the majority, citizens are easily convinced to migrate to developed countries in search of a better life (Van Heelsum, 2017: 2141). The number of asylum requests from Eritreans from total asylum applications has seen an increase from 10% in 2014 to 15% in the first half of 2017 (Sterckx et al., 2018). The oppression and human rights violation in Eritrea enable Eritrean migrants to access 'long-term protection' in the Netherlands. However, trends have shown a decline in the provision of residence permits for asylum seekers from 87% in 2016 to less than 30% in 2017 (Ministry of Justice and Security in Pharos Netherlands 2019: 1). Despite this trend, the influx of migrants to the Netherlands remains unabated. Although, generally, migrants try to adapt to the living condition and lifestyle in developed countries such as the Netherlands, living as a migrant brings its own share of challenges, especially when dealing with an unprecedented global crisis like the COVID-19 pandemic.

4.3 Intersecting Challenges of Eritrean and Ethiopian Migrants During COVID-19

The situation of Eritrean and Ethiopian migrants in the Netherlands is perhaps no different as compared to other migrant groups here in the Netherlands or elsewhere. Although many are affected by the coronavirus, vulnerable groups such as migrants and those engaged in precarious work are more exposed to its plight. Della Rosa and Goldstein (2020) contest the generalized reports and narratives that claim COVID-19 as an equalizer of a previously unbalanced structure. They argue that the virus itself does not discriminate between rich and poor people or nations and has proven to be an enigma that even the most economically powerful nations could not control. However, those in lower economic conditions, marginalized groups, and migrants continue to bear the worst effect of the pandemic (Della Rosa & Goldstein, 2020: 1). As stated by Norman (2020):

"...although many people all over the world are now riding the same storm, they are doing so in very different boats" (Norman, 2020: 1).

All the findings of our research pointed to the intersection of different identities that exacerbate Eritrean and Ethiopian migrants' experience with COVID-19. Whether it is children, recently arrived migrants, women, single mothers, all faced circumstances that intersected to influence and, in most cases, worsen their experience with the pandemic.

One of our key informants, who is a co-founder of the helpdesk, explained the situation during the lockdown, when hospitals discouraged in-person appointments unless for serious health cases as follows:

"Language is a major barrier for recent migrants, they might speak enough Dutch or English to make an appointment but not to explain their issues in detail, which hinders them from getting an appointment at the hospital. In severe health cases, volunteers at the helpdesk called the general practitioners themselves to explain and obtain an appointment" [Helpdesk, Co-founder].

Our research participants have one unifying factor, that is, their migrant identity in the Netherlands. This migrant status of individuals is a prominent factor exposing them to various vulnerabilities, commonly reflected by majority of our research participants. However, a key point worth noting is, even within this migrant population, experiences differ. Some coped relatively better while others suffered more based on factors such as gender, age, marital status, health status, means of livelihood, educational status, knowledge of Dutch and English language, and other personal characteristics.

While this pandemic has become a uniting factor for some, it has created a clear point of demarcation, among others. For instance, many nations have closed borders to non-citizens. "Citizenship appears to have resurfaced as the ultimate marker of belonging and solidarity" (Triandafyllidou, 2020: 261). However, this does not mean citizens of a country are less likely to be infected by the virus or carry it to their countries of origin, it just means national solidarity transcends protection from the virus. "In other words, states weigh their obligation towards solidarity and protection of citizens above the risk that they may be carrying the virus" (Triandafyllidou, 2020: 261). This pleads questions like, what about those who do not reside in their home countries and those who cannot travel back to their countries? What about those who are striving to make a home outside of their home? Those who are labeled 'migrants' and treated as second class citizens? Who stands for humanity outside the confinement of national boundaries and citizenship, especially during an unforeseen catastrophe such as the COVID-19 pandemic?

The prominent challenges faced by the Eritrean and Ethiopia community are presented in the following sub-sections.

4.3.1 The Downside of Staying Inside – Fear, Anxiety and Confusion

The NPI measures such as staying home and physical distancing, which are undertaken to curb pandemics have dire psychological consequences, which include anxiety, stress, and depression. These impacts are not only felt during the pandemic but might cause prolonged complications for a person's mental health (Rauschenberg et al., 2020: 5). Impacts, such as depression, anxiety, and panic attacks, were also reported by interviewed Eritrean and Ethiopian migrants. A female respondent who is an asylum seeker living in Rijswijk camp for the last 3 years shares her concern about social distancing as follows:

> "I worry because I live in a camp. I share a living room, a bathroom, and a kitchen with 7 roommates while 3 of us stay in one room. Since it is a shared living space, I am afraid, even if I apply all protective measures, my roommates might still expose me to the virus" [Female, 29, Asylum Seeker].

Kluge et al. (2020) also argue that refugee camps are particularly prone to expose their residents to virus infection. "These camps usually provide inadequate and overcrowded living arrangements that present a severe health risk to inhabitants and host populations" (Kluge et al., 2020: 1238).

The psychological impacts of isolation and distancing are potentially worse among migrants adjusting to a foreign land, away from loved ones. Ironically, when asked about what worried them, interviewees reported greater anxiety about the conditions back in their country of origin than about the situation in the Netherlands. They worried about the vulnerable health systems, poor livelihoods, and lack of social safety nets in Eritrea and Ethiopia. A male respondent shares his concerns about the conditions in Ethiopia:

> "I am worried about the situation back home because the economic condition depends on people's daily labor. People will not follow distancing and staying home measures because they cannot afford to do so. The recent political instability in Ethiopia is also adding fuel to the fire" [Male, 50, Engineer].

As discussed in Ayenew and Pandey (2020), Ethiopia has a "…very low health care workforce density of about 0.96 for every 1000 population". This alarming statistics "coupled with shortage of hospitals, mass use of public transportation, shortage of sanitation materials including water, hiding suspected cases, lack of personal protective equipment for health care providers, presence of immune-compromised people are among the major driving factors making Ethiopia one of the challenged developing country in facing this unprecedented COVID-19 spread" (Ayenew & Pandey, 2020:1).

In relation to the weak health system in Ethiopia, a 30-year-old interviewee from Eritrea shared his heightened worry when he learnt about two people testing positive in *Mai-Ayni* refugee camp in Ethiopia, where his family is currently living. As explained in Kluge et al. (2020), in most cases, refugee camps do not provide essential survival needs such as clean water, soap and are marked by a lack of medical professionals and insufficient access to vital health-related information. This is especially worse in developing countries, like Ethiopia, where *Mai-Ayni* camp is located.

Here, we can observe the layers of struggle faced by migrants. Hearts and thoughts are torn between home and the place where they currently reside. Even if migrants feel they have managed to get themselves to better living conditions in developed countries, their hearts and thoughts remain with their families back home whom they perceive are in difficult conditions. The inability to travel and see loved ones and continued internet shutdown in Ethiopia prevents migrants from contacting their families.

4.3.2 Heterogeneity of Migrants: Financial Shock Felt Differently by Different Migrants

Financially, as may be expected, the pandemic has impacted migrants differently depending on their field of occupation. Interviewed migrants are engaged in various sectors including service industry (waiter, food delivery, taxi drivers), caretaking for the elderly, business owners, and some employed in professional institutions.

Businesses such as freelancers, taxi drivers with own vehicles, and restaurant owners, experienced the worst financial effect of the pandemic. The Dutch government provided a onetime payment of 4000 Euros to business owners in March 2020. For those who lost their jobs, the government provided unemployment benefit of 1050 for singles and 1500 for families. Businesses lost significant portion of their earnings and were hence forced to dip into their savings. For instance, a 46-year-old male taxi driver reported a decrease in his monthly income from 4500 Euros to an unemployment benefit of 1500 Euros. This shows a significant plummet in his income, which forced him to make various adjustments on expenses.

On the other hand, as reflected from the discussion of migrants from Eritrea who work in food deliveries, those working in food delivery experienced an increase in income, which could be due to the preference to order online than going out during lockdown. The relatively low language, education and skill requirements in the sector seems to attract high participation of migrants, especially recent migrants. However, this comes at the cost of exposure to health risks as more hours spent outside increase the chances of infection, which shows the precarious and risky nature of the delivery job.

This is supported by Huijsmans (2020) who argues that those who work in delivery jobs acknowledge the risks brought about by moving across different restaurants to pick up food and going to different locations/homes to deliver.

> "Especially in places where one knows things have been touched a lot by many different people … and you have to touch that button or hold that door handle, "you know there is something wrong, but you have to [do it]", one Deliveroo rider remarked" (Huijsmans, 2020)

As stated in Gelatt (2020), people's occupation directly influences their risk of contracting COVID-19. While many companies, institutions and professionals have opted to go virtual, service sector jobs such as deliveries, caretaking, cleaning, and other service-related jobs, which are mainly occupied by migrants continue to be performed physically, hence increasing migrant's risk to the virus.

Similarly, Siegmann (2020) brings to the forefront the underrated work performed by low wage workers engaged in food delivery and care work. The work of those participating in the service industry can be classified as "essential work". Though essential, these jobs are also considered "precarious" because they present a risk to the workers. "Thus, while symbolic and literal applause for essential workers reveals a level of cognizance of their importance, in fact, the coronavirus crisis even aggravates these workers' precarity" (Siegmann, 2020: 1).

In the UK, a survey of 2108 adults revealed high interest among respondents to stay isolated, if they could (Atchison et al., 2020). This shows an inability to cope with the consequences, rather than a lack of interest that hinders many individuals from applying preventive approach such as staying home. According to this study, "those with the lowest household income were six times less likely to be able to work from home and three times less likely to be able to self-isolate". They further argued that "[t]he ability to adapt and comply with certain NPIs is lower in the most economically disadvantaged in society" (Atchison et al., 2020: 2).

Some found themselves in a more difficult situation as they were laid off from their jobs. A 30-year-old male from Eritrea, who was a coffee machine operator in a restaurant, explained his vulnerability since his contract has been terminated due to financial loss to the employing restaurant during the lockdown. He further explained the inadequacy of 375 Euros compensation to cover his expenditure. On the other hand, a 50-year-old male, working as engineer, discussed that both he and his wife were granted a 10-day care leave that can be used to care of children or other reasons. He explained this was particularly helpful to manage his workload and homeschooling of kids during the lockdown. This clearly shows the inequality within migrants. As recent Eritrean and Ethiopian migrants are not employed in secure jobs, this delineates the difference in the financial effect of the pandemic among migrants.

The mainstream narrative states that those with already existing medical conditions are more threatened by COVID-19. Nevertheless, poor health conditions intersect with poor economic conditions and many people fall under both categories, exacerbating their exposure to COVID-19.

> "The COVID-19 pandemic has illuminated the stratification of society in every nation-state it has touched. The pandemic has unmasked the hidden systems of inequality that are lost in the mundanity of everyday life fracturing the veneer of capitalist meritocratic society" (Nolan, 2020: 1).

A representative of Pharos stated that most recent Eritrean migrants do not come from the cities. They come from rural areas, which increases the likelihood of being uneducated. This makes it difficult for them to easily integrate within the Dutch environment. She discusses "…the amount of paperwork, I can't imagine how you deal with that when you have not learned how to read or write". This demonstrates the interplay between lack of education and the migrant status of individuals in excluding them from the system and any financial or health benefits.

4.3.3 Risking Health for Social Capital

According to the discussion by a representative from Pharos, it was difficult for Eritreans to follow the rules of staying home. Their hesitation to physically visit people was interpreted as 'uncaring' character, due to cultural expectations, which reserve phone calls and other ways of communication for distant family and friends. Within these cultures, those who truly care about other's wellbeing, always take the time to visit. As a result, many braved into the houses of loved once, while bearing the risk.

Cruwys et al. (2020) "social identity model of health risk-taking" explains this finding. According to his model, which is based on the social identity theory, humans have two separate but also interrelated means of locating sense of identity. On a more personal basis, we identify ourselves as individuals with a unique identity. We also conform to social identities that are part of a bigger social structure, emanating from various groups such as religion. We usually hold these groups near and dear to our hearts while also taking pride to be associated with them.

> "Through their capacity to transform psychology and behavior, group processes fundamentally structure our perception of safety versus vulnerability. … potential threats arising from ingroup members – particularly those with whom we share a strong social identity – will be perceived as less risky, and inspire greater risk-taking behavior" (Cruwys et al., 2020: 585).

For these migrants, who are connected by their migrant status in a foreign land, their country of origin (Eritrea/Ethiopia) and/or religious backgrounds (Christianity/Islam) constitute essential parts of their identity. These forms of social identities influence the decision to implement COVID-19 preventive measures or not. In some cases, they made a deliberate choice to conform to social expectations while disregarding public health regulations to spend time with friends and loved ones during the holiday seasons.

As per the discussion with the co-founder of the helpdesk, the number of migrant infections skyrocketed during Easter and Eid holidays in April and May 2020 because people were gathering for celebrations. Even if the church was closed, people still risked the fine of 390 Euros and kneeled in front of church gates to get God's blessings. A belief in God's protection seems to transcend over the fear of the pandemic. This explains the rise in the number of infections among Eritrean and Ethiopian migrants during the festivities of Easter and Eid-Al-Adha.

4.3.4 Experience of Migrant Women with COVID-19

Based on our interviewees, we noticed that in some households, particularly those who have arrived in the Netherlands in the recent past, household chores are typically left for women. Some women, especially stay-at-home mothers are heavily reliant on their husbands/partners to obtain any information, including information on COVID-19. For example, a 36-year-old female informed that she is completely dependent and subservient to her husband. She relied on him to manage household income, make doctor's appointment, and even obtain information about COVID-19. Furthermore, some interviewed Eritrean women were victims of domestic abuse by their husbands/partners. For these women, verbal abuse and heightened vulnerability have become a day-to-day phenomena during the lockdown.

As confirmed by a representative from the helpdesk, domestic violence cases have always been an issue among the Eritrean community, but the numbers increased during the lockdown. During the lockdown, physical and verbal abuse were happening in the presence of kids. When the helpdesk reported cases of violence to the concerned body, investigators refused to help because they did not feel safe to go to the reported households due to the risk of COVID-19 infections.

Sieffien et al. (2020) also argue that increase in domestic violence is one of the ramifications of staying home, especially for immigrants and asylum seekers. "High density, close-quartered living conditions, debilitating poverty … could be risk factors more commonly found in this population" (Sieffien et al., 2020: 2).

Our research findings also indicate the disproportionate share of household chores borne by mothers during the lockdown. This is especially worse for single mothers who need to take care of their kids while doing household chores and working from home. Lewis in Power (2020) makes an interesting observation that "school closures and household isolation are moving the work of caring for children from the paid economy-nurseries, schools, babysitters- to the unpaid one" (Power, 2020: 68). As stated by the United Nations (UN) in Power (2020), this care work usually falls on women, which worsens their situation.

In addition to performing household chores, the responsibility of schooling kids and taking care of other family members in the household is left for women, even when they are performing paid work. This is "because of the persistence of traditional gender roles and partly because of the structure of women's economic participation, which is more likely to be part-time, flexible, and less remunerative" (UN in Power, 2020: 69).

A 28-year-old female widow who is also a single mother of two children expressed her worries as follows:

> "I panicked every time a post came to my mailbox because I did not understand what it meant. Under normal circumstances, I used to take letters to organizations such as Dutch Refugee Council (Vluchtelingenwerk Nederland/VWN), but they were closed, which made me too stressed because I did not want to end up in debt" [Female, 28, Unemployed].

She continued expressing her frustration with the hospitals when her son fell sick:

"My son was very ill, but I could not reach the general practitioner. They were asking me too many questions, which I did not understand so I hang up the phone and just cried. Afterwards I heard about the helpdesk and was able to get an appointment with their help" [Female, 28, Unemployed].

4.3.5 Family Status Determining Experience of Children with COVID-19

Economic conditions and educational background intersected with the migrant status of parents and families to influence children's experience with online education. Language is one of the major barriers for recent migrants. Most of the recent Eritrean and some Ethiopian migrant parents do not have a good command of both Dutch and English languages, which prevents them from understanding emails and communications sent from schools.

As explained by the co-founder of the helpdesk, many less-educated migrants were in panic as they did not understand the information shared by schools. Online schooling was a big problem since they did not know how to support or guide their children. "There were many who could not differentiate between a password and username to login to the system of the schools". Even after they managed to log in with the help of volunteers at the helpdesk (through video calls), they could not understand the instructions to support their kids, which resulted in some children falling behind and attending mandatory summer schools to catch up.

Findings also show that in some households, there was a mismatch between the number of computers and the number of children in the household. As explained by the helpdesk representative, "Schools were alarmed because the majority of migrant kids were not participating in the online education system; these schools were calling parents and the helpdesk to reach out the children".

Similar observations were made by research conducted in Turkey and Germany. The online schooling system introduced due to COVID-19 provided a blanket solution to all residing in these countries, but failed to consider the specific conditions of marginalized groups, specifically migrants. The problems faced by migrant families and children, especially refugees include the inability to access the online education system arising from lack of gadgets such as computers, tablets, or smartphones in the household. Under these conditions, communication between the schoolteachers and the students was completely halted (see also Gómez & Andrés, 2021). For example, in Turkey, around half of refugee children were unable to conduct homeschooling during the lockdown (Kollender &Nimer, 2020: 5).

In contrast, when speaking to educated parents, results showed that migrant parents who have been trained in the Dutch education system from their bachelors or earlier found it relatively easy to support their kids with schoolwork. A father of four daughters, who has lived in the Netherlands for longer than 30 years discusses:

"It was a time when I was able to regularly check in on my children's education" [Male, 46, Taxi driver].

For those who were educated in their countries of origin, they found the Dutch method of education quite puzzling. As described by one interviewee:

"I was almost depressed because it was too much to handle; juggling my work, parenting my kids that were home 24/7 and supporting their online education while dealing with the stress of a pandemic" [Female, 38, Consultancy firm owner]

Similarly, a mother of two children who had to manage her online accounting work with homeschooling kids expressed her dismay:

"The kids were restless at home, not as disciplined as they would be in school. The home setup is obviously different from school, so they were acting as if they were on a break and unable to take schoolwork seriously" [Female, 38, Accountant].

Others complained about the loads of emails crowding their inboxes daily with specific instructions. Narrating her experience:

"Even for me as an educated person, it was difficult to understand the teaching system. When I tried to explain what I understood, the kids refused my methods, they only accepted what the teacher says as if it is a scripture from the bible" [Female, 38, Consultancy firm owner].

Issues related to living conditions were also discussed by some families living in small apartments that felt suffocated when the whole family had to stay in for a long period. Kids were reported to fight more often, disrupting the household.

When asked about the quality of education, most parents expressed their concern with online education. It was a new system, which many did not fully comprehend and therefore could not support their children. Additionally, the kids were distracted because the setting was different from the formal school.

On a positive note, some parents appreciated the close monitoring of students by the schools.

"I am very impressed with the teachers' and the schools' commitment. They were following up with each student, making individual phone calls, and providing reading materials. The online classes were carried out well" [Male, 46, Taxi driver].

The above discussion shows a varied experience of migrant kids with online education depending on factors including economic conditions and educational background of parents.

4.4 Conclusions and Policy Implications

COVID-19 experiences are differentiated based on numerous identities and social factors. Our research findings have revealed the intersecting axes of vulnerabilities that have exacerbated the experience of Eritrean and Ethiopian migrants with the COVID-19 pandemic in The Hague. Missing family members, worrying about loved ones back home, employment in precarious and unstable jobs, coupled with fear for one's life are some of the tough realities of the life of these migrants. Isolation and distancing, that appear as easy concepts to grasp as well as implement on the surface, pose a challenge for migrants who live in crowded living conditions.

Language is a major barrier that hinders recent migrants from obtaining crucial health-related information and other social benefits. When looking into the financial impact of the pandemic, the migrant identity of individuals intersected with type of employment in determining the financial effect of the COVID-19 crisis. Gender identity also intersected with the migrant status of women exacerbating their experience with the pandemic, which is related to the gender-based division of labor in the countries of origin that leave most or all household chores to women.

The intersection of economic conditions and educational status also impacted children's online education experience during the lockdown. Parents with low level of education and income faced difficulty to support their children with online schooling. This ensued significant negative implications and resulted in the requirement of some children to attend summer school.

A keen approach to deal with infectious disease such as COVID-19 is an all-inclusive one that does not discriminate based on economic or social status. Excluding migrants from awareness creation, health services and information does not only have negative consequence for migrants themselves but also 'undermines the effectiveness of relevant public health efforts' (Guadagno, 2020: 13).

Our findings call for COVID-19 preventive measures to include the needs of migrants especially by institutions working with migrants in the Netherlands such as municipalities (*Gemeentes*), Dutch Refugee Council (*Vluchtelingenwerk Nederland*), Pharos and others. These organizations could make use of insights and inputs from migrants themselves to design and implement projects targeting migrants. Moreover, these organizations could explore the potential of settled migrants by using them as cultural mediators (*sleutelpersonen*) to serve as a bridge between new migrants and the organization's staff, which are mostly local, Dutch.

To curtail the language barrier, translated and visualized information can be provided on preferred media (for example, Facebook and WhatsApp). Findings reveal that many Ethiopian and Eritrean migrants exhibit strong religious identities, thus religious institutions and communities can be used as platforms to disseminate information.

Another important field to address is online education. Government support should prioritize low-income families especially when implementing a fully online education system. Special assistance should also be provided by schools for migrant kids with less-educated parents.

Pandemics, as in the case of COVID-19, call for solidarity by all, especially for groups who share many similarities such as migrants from the same country of origin. Settled Ethiopian and Eritrean migrants should also collaborate to support recent ones by serving as cultural mediators. Having been in the position of new migrants themselves, they fully comprehend and can even predict specific needs of recent migrants. They should be reminded of their fellow community members who have not yet acculturated fully to the community and hence prone to confusion in a new land. Interventions involving migrants themselves to support other migrants can complement efforts made by the Dutch government.

In short, our findings suggest that a blanket solution to all is not only impractical but could also exacerbate the vulnerability of marginalized groups, such as migrants. Hence, specific problems unique to migrants require specific and contextual solutions.

References

Atchison, C. J., Bowman, L., Vrinten, C., Redd, R., Pristera, P., Eaton, J. W., & Ward, H. (2020). Perceptions and behavioural responses of the general public during the COVID-19 pandemic: A cross-sectional survey of UK adults. *BMJ Open, 11*(1), e043577.

Ayenew, B., & Pandey, D. (2020). Challenges and opportunities to tackle COVID-19 spread in Ethiopia. *Journal of PeerScientist, 2*(2), e1000014.

Cruwys, T., Stevens, M., & Greenaway, K. H. (2020). A social identity perspective on COVID-19: Health risk is affected by shared group membership. *British Journal of Social Psychology, 59*(3), 584–593.

Della Rosa, A., & Goldstein, A. (2020). What does COVID-19 distract us from? A migration studies perspective on the inequities of attention. *Social Anthropology, 28*(2), 257–259.

Gelatt, J. (2020). *Immigrant workers: Vital to the US COVID-19 response, disproportionately vulnerable*. Migration Policy Institute. Available at: https://www.migrationpolicy.org/research/immigrant-workers-us-covid-19-response#:~:text=March%202020-,Immigrant%20Workers%3A%20Vital%20to%20the,COVID%2D19%20Response%2C%20Disproportionately%20Vulnerable&text=Six%20million%20immigrant%20workers%20are,during%20the%20COVID%2D19%20pandemic. Accessed on 20 Mar 2021.

Gómez, G. M., & Andrés Uzín, G. J. P. (2021). Effects of COVID-19 on education and schools' reopening in Latin America. In E. Papyrakis (Ed.), *Covid-19 and international development*. Springer.

Guadagno, L. (2020). *Migrants and the COVID-19 pandemic: An initial analysis* (Migration research series no 60). International Organization for Migration.

Huijsmans, H. (2020). *Restaurants are empty, but the work continues: Freelance food delivery in times of COVID-19*. https://issblog.nl/2020/05/18/COVID-19-restaurants-are-empty-but-the-work-continues-freelance-food-delivery-in-times-of-COVID-19-by-roy-huijsmans/. Accessed on 17 Mar 2021.

Kluge, H. H. P., Jakab, Z., Bartovic, J., D'Anna, V., & Severoni, S. (2020). Refugee and migrant health in the COVID-19 response. *The Lancet, 395*(10232), 1237–1239.

Kollender, E., & Nimer, M. (2020). *Long-term exclusionary effects of COVID-19 for refugee children in the German and Turkish education systems: A comparative perspective*, IPC–MERCATOR Policy Brief, July 2020.

Murshed, S. M. (2021). Consequences of the Covid-19 pandemic for economic inequality. In E. Papyrakis (Ed.), *Covid-19 and international development*. Springer.

Nolan, R. (2020). 'We are all in this together!' COVID-19 and the lie of solidarity. *Irish Journal of Sociology, 29*(1), 1-2-106.

Norman, J. (2020). *Gender and COVID-19: The immediate impact the crisis is having on women*, British Policy and Politics at the London School of Economics (LSE), Blog entry, April 2020. Available at: https://blogs.lse.ac.uk/politicsandpolicy/gender-and-covid19/#:~:text=April%2023rd%2C%202020-,Gender%20and%20Covid%2D19%3A%20the%20immediate%20impact%20the,crisis%20is%20having%20on%20women&text=Women%20do%2C%20on%20average%2C%2060,less%20time%20for%20paid%20work. Accessed on 20 Mar 2021.

Ong'ayo, A. (2010). Ethiopian organisations in the Netherlands. In A. Warnecke (Ed.), *Towards a comparative assessment of factors determining diaspora intervention in conflict settings: Somali and Ethiopian diaspora Organisations in Europe*. Bonn International Centre for Conversion.

Pharos. (2019). *Eritrese vluchtelingen*. https://www.pharos.nl/factsheets/eritresevluchtelingen/. Accessed on 14 Nov 2020.

Pharos. (2020). *English*. https://www.pharos.nl/english/. Accessed on 14 Sept 2020.

Power, K. (2020). The COVID-19 pandemic has increased the care burden of women and families. *Sustainability: Science, Practice and Policy, 16*(1), 67–73.

Rauschenberg, C., Schick, A., Goetzl, C., Roehr, S., Riedel-Heller, S. G., Koppe, G., Durstewitz, D., Krumm, S., & Reininghaus, U. (2020). Social isolation, mental health and use of digital interventions in youth during the COVID-19 pandemic: A nationally representative survey. *European Psychiatry, 64*(1), e20. https://doi.org/10.1192/j.eurpsy.2021.17

Sieffien, W., Law, S., & Andermann, L. (2020). *Immigrant and refugee mental health during the COVID-19 pandemic: Additional key considerations*. Canadian Family Physician, Mississauga, Canada. Available at: https://www.cfp.ca/news/2020/06/23/06-23-1. Accessed on 21 Mar 2021.

Siegmann, K. A. (2020). From clapping for essential workers to revaluing them. *Global Labour Column, 339*, 1–2.49.

Sterckx, L., Fessehazion, M. & Teklemariam B. E. (2018). *Eritrean asylum status holders in the Netherlands*. The Netherlands Institute for Social Research, the Hague, the Netherlands. Available at: https://www.scp.nl/english/Publications/Summaries_by_year/Summaries_2018/Eritrean_asylum_status_holders_in_the_Netherlands. Accessed on 21 Mar 2021.

Triandafyllidou, A. (2020). Commentary: Spaces of solidarity and spaces of exception at the times of Covid-19. *International Migration, 58*(3), 261.

Van Heelsum, A. (2017). Aspirations and frustrations: Experiences of recent refugees in the Netherlands. *Ethnic and Racial Studies, 40*(13), 2137–2150.

Vieira, C. M., Franco, O. H., Restrepo, C. G., & Abel, T. (2020). COVID-19: The forgotten priorities of the pandemic. *Maturitas, 136*, 138–141.

Chapter 5
Consequences of the Covid-19 Pandemic for Economic Inequality

Syed Mansoob Murshed

Abstract The impact of the COVID-19 pandemic is likely to increase various forms of economic inequality in wealth and income. This is because the income of the poor was adversely affected more, both because of the already present technology driven trends in unskilled labour substitution, but also because the types of employment that the world's poor engage in was most severely disrupted by COVID-19, and the subsequent public health response. This is in contrast to medieval pandemics, which tended to increase the wage-rental ratio. Certain countervailing income and job protection schemes can help, but it is mainly a short-term palliative. Population weighted international inequality has also increased. Unless checked, further increases in inequality will strengthen recent trends in illiberal, populist, governance.

5.1 Introduction

The Covid-19 pandemic has afflicted humanity since the beginning of 2020, and no end seems in sight at the time of writing, despite progress in developing vaccines and treatments for the disease. In addition to the human costs of lost lives, and the adverse long-term health effects for many who have survived the illness, there are also major economic costs. The pandemic engendered a major downturn in most economies, resulting in a dramatic decline in output of a much greater magnitude than is usual with most recessions. This is because of a decline in aggregate demand, in addition to the supply shock engendered by the pandemic. Economic recessions

---

S. M. Murshed (✉)
International Institute of Social Studies (ISS), Erasmus University of Rotterdam (EUR), The Hague, The Netherlands

Centre for Financial and Corporate Integrity (CFCI), Coventry University, Coventry, UK
e-mail: Murshed@iss.nl

© The Author(s), under exclusive license to Springer Nature Switzerland AG 2022
E. Papyrakis (ed.), *COVID-19 and International Development*,
https://doi.org/10.1007/978-3-030-82339-9_5

are normally associated with increasing inequality, as well as poverty,[1] because the poorer segments of society are disproportionately affected by economic contraction in terms of the loss of income and employment (see also Demena et al., 2021 and Romanello, 2021). The more impoverished are also more greatly affected by the disease burden, because they are less able to take precautionary measures such as social distancing (Fantu et al., 2021; Wagner, 2021). Thus, the adverse impact of the pandemic both on the health and economic well-being of the more disadvantaged in society will be greater, implying rising inequality. There appears, therefore, not much to debate about the direction of the consequences of the Covid-19 on economic inequality. The mechanisms through which this inequality increasing effect works, however, merit attention, and that is precisely what this work essays to do. The measurement of the consequences of the pandemic will require revision for years to come.

Historically, protracted pandemics and wars can often be harbingers of systemic change, including the distribution of income, and this is what is discussed in the next section. In the section that follows, we work out the simple macroeconomics of pandemics on output, employment and inequality along with a summary of the scant (mainly conjectural) empirical evidence on inequality effects that currently exist. This is followed by a section on the possible effects of the pandemic on global inequality and the globalisation-inequality nexus. The last section concludes with a cautionary note on the consequences of yet more increased inequality for democratic governance.

Before proceeding further, some stylised facts about inequality need mention. A degree of economic inequality is inevitable in any society, reflecting the price paid for the incentivization of risk and work effort, but inequality that is systemically perpetuated, preventing certain groups in society and the under-privileged from rising, is described as *inequality of opportunity*. This form of inequality is both undesirable and economically inefficient (Stiglitz, 2012; Roemer, 1998). Inequality has been rising all over the world and has mainly been ascribed to labour saving technical progress (Dabla-Norris et al., 2015) and financial globalisation (Furceri et al., 2017). The former leads to real wage compression for unskilled labour, as well as precarious employment; the latter causes further increases in the capital share of national product, exacerbating the weakened bargaining power of labour over capital. Inequalities in income are the most common inequality metric, but inequalities in wealth are far greater, and the ownership of wealth is far more concentrated compared to income.[2] The chief concerns with these developments are to do with the income and wealth share of the richest 1% of the global population, whose income

---

[1] This work will concern itself with inequality and not poverty. Increases in poverty due to the pandemic merit independent analyses; with falling output and rising inequality, poverty is bound to increase.

[2] According to Piketty (2014), the accelerating trend in inequality mainly stems from wealth inequality. There is a tendency for wealth to national income ratios to increase since the 1970s; wealth, whose ownership is more concentrated than income, multiplies faster than wage income creating an ever widening gap between capital and labour, the biggest source of inequality.

and wealth is rising faster compared to all groups, including the relatively less rich. Inequality is usually measured using indices, the most common among which is the GINI[3] coefficient. Another point about inequality which needs to be borne in mind is the fact that as average income increases it is feasible to have greater inequality, as the distance between subsistence (or socially acceptable minimum living standards) and average income grows: this is referred to as inequality possibility (Milanovic, 2016) or the extraction rate (Scheidel, 2017).

5.2 Pandemics, Protracted Violence and Inequality

Earlier pandemics have sometimes produced, or set in train, transformations in the sense described by Polyani (1944). The Black Death (*magna pestilencia*) of circa 1348-50, heralded the demise of the feudal economic system (Scheidel, 2017; Routt, 2008), a process which was to last three centuries, ending with the cataclysmic Thirty Years War of 1618-48 (Hobsbawm, 1954). The decimation of the population following the Black Death[4] made land more abundant relative to labour, increasing the wage-rental ratio. This had the impact of lowering inequality, a tendency moderated by the falling price of grain in the late fourteenth century. It also altered the composition of output, leading to less grain production, increased animal husbandry, and the pattern of Europe-wide trade in woollen products, halting the hitherto flourishing Silk Route trade (Routt, 2008). There was also political turmoil, with peasant revolts in Belgium prior to the Black Death, in England in 1381, and in Germany in the early sixteenth century (Scheidel, 2017). These may have had some levelling effects. The feudal system, however, lingered till the end of the Thirty Years War (Hobsbawm, 1954), which also produced intermittent epidemics. Hobsbawm (1954) characterized the seventeenth century as a period of crisis in Europe. The conclusion of that catastrophe radically altered the global economic system in many ways, chief of which was the decline of Venetian Mediterranean trade. Citing detailed evidence from the devastated German city of Augsburg, Scheidel (2017) outlines the 'equalizing' consequences of physical devastation, particularly through the proscription of the richest segments of society.

Voightländer and Voth (2013) point to wars and war-related epidemics from the Black Death up to the eighteenth century arguing that the decrease in population caused an increase in per-capita income, as the stock and productivity of land was left unaffected. This provided rulers with a taxable surplus, which they invested in armies and armaments, producing even more war.

---

[3] The GINI coefficient is the sum of differences of all incomes or income groups from the mean or average income. It is usually computed to range from 0 to 100, with the former implying perfect equality, and the latter perfect inequality, such that increases in the GINI imply more inequality.
[4] The Black Death may have killed up to half the population in England and a third of the Italian population; Scheidel (2017).

Using an apocalyptic, Biblical, analogy Scheidel (2017) speaks of 'four horsemen' which have acted as the great (economic) levellers throughout history: total war necessitating mass mobilization (not short-lived wars fought by professional armies), state collapse, pandemics and political revolutions. All four of these phenomena, are inter-related, and endogenous to each other. Indeed, Scheidel (2017) goes far enough to suggest that these four violent forces are the only ones that engender sustained decline in inequality. He begins by drawing attention to the Roman empire of classical antiquity where the concentration of wealth at the top mirrors much of the present-day wealth holdings of the top 1%. The decline of the (Western) Roman empire had a strong equalizing effect on economic inequality in Europe until the high Middle ages, preceding the Black Death of the fourteenth century, which halted the trend increase in inequality for a century and a half. Similar, developments can be discerned for Asia, with some differences in timing. He argues, in accordance with other scholars, that the next biggest event that reduced inequality were the two world wars of the twentieth century, commencing in 1914, and continuing until the 1970s/80s due to welfarist and egalitarian policies pursued in the aftermath of the Second World War. This period of increasing equalization (1914–80) is also the period of declining wealth to GDP ratios (and falling rental to growth ratios) highlighted by Piketty (2014). It also suggests that the two historical episodes of economic globalisation (1870-1914 and 1980 to the present) demonstrate Kuznet cycles; inequality first rises and then falls as globalisation diminishes (Milanovic, 2016).[5] We have, however, not yet witnessed a discernible end to our current phase of globalisation, although there may be indications of de-globalisation (van Bergeijk, 2019).

Barro et al. (2020) study the economic impact of the great Spanish influenza pandemic of 1918-20, an event on which we have reliable data. That pandemic killed off about 2% of the global population (with the greatest mortality rate as a proportion of the population being in India), and on average GDP declined by 6% per country. The Spanish influenza pandemic may have reinforced the equalization forces set about the cataclysmic events of the Great War of 1914-18.

The present pandemic's impact on wealth and income inequality is yet to emerge, but what must be borne in mind is that in present-day production labour is more substitutable than ever before due to artificial intelligence and robotisation, providing scope for the capital-labour ratio to rise. The wage-rental ratio is, therefore, unlikely to rise in the wake of the current pandemic, suggesting that channel of equalisation is not in force unlike with medieval plagues.

---

[5] Milanovic (2016), in contrast to Scheidel (2017), argues that income inequality, especially in the UK, began falling well before 1914.

5.3 Macroeconomics of the COVID-19 Pandemic and Inequality Trends

The current pandemic, and its associated lockdowns, has resulted in both a negative supply shock as supply chains are disrupted, and a negative demand shock as disposable incomes diminish. For households, this is chiefly due to the diminution in labour supply, either because of the reduced demand for labour (involuntary unemployment) or the voluntary withdrawal of labour engendered by the fear of infection. The reduction in labour demand is likely to be greatest for the provision of personal services, which is the category most pervasive among precariously employed and low-income workers. By contrast, higher income knowledge workers are hardly affected because of their ability to work remotely. Adverse aggregate demand (consumption) effects of unemployment amongst workers who cannot work from home can be mitigated by employment and income support schemes, and many such policies have been enacted.

It is worthwhile considering a simple (Keynesian) macroeconomic model with heterogenous households in terms of their income, assets and lockdown employment prospects in the spirit of the model constructed by Kaplan et al. (2020). This is described below:

$$Y(R) = \sum_{i=1}^{n} C_i\left(Y_i^D\right) + I(r,R) + G + T \quad (5.1)$$

Where

$$Y_i^D = (1-t)w_i L_i(R) + B_i \quad (5.2)$$

In the above Y stands for aggregate supply or national income, which is negatively related to the intensity of infection, R. National income is postulated to accrue to households. We envisage a variety of household groups ranging from 1 to n, arrayed in order of their income and wealth. C stands for consumption, I for investment which is a negative function of both interest rates, r and the intensity of infection, R, G stands for government expenditure, and T for the external trade balance (exports minus imports). As indicated, households are heterogenous or unequal in terms of their disposable income, Y_i^D, which is defined in (2) as being equal to wage income, w from employment, L, less a proportionate tax, t and liquid assets B available to household, (i). Poorer households will receive lower wage rates, and are more likely to be laid off due to lockdown responses to the rising infection rates. They are likely to have fewer liquid assets (savings) at their disposal for emergencies; indeed, this income component will be negative for indebted households.

Totally differentiating (5.2) with respect to the infection rate, R, setting initial $t = 0$, we find:

$$\frac{dY_i^D}{dR} = L_i \frac{dw_i}{dR} + w_i \frac{\partial L_i}{\partial R} dL_i + \frac{dB_i}{dR} \qquad (5.3)$$

The terms on the right-hand side of (5.3) are likely to be negative the poorer the household. For more affluent and highly skilled workers who can work from home, the first two terms on the right-hand side of (5.3) will be unaffected, and for some high skilled occupations may even be positive. The second term on the right-hand side of (5.3) may be negative on account of the voluntary withdrawal of labour by some workers who fear (and derive disutility from) infection at the work-place. The liquid asset position of better remunerated workers may also increase, as they consume less due to lockdown. The adverse effects on low-income, unskilled workers unable to work from home may be mitigated by employment and income support policies. In short, inequality based on household income and consumption will increase, as soon as the income/employment protection schemes cease. Also, the functional distribution of income will move in favour of the more skilled and those able to work from home, given the heightened demand for skills related to information technology and artificial intelligence aimed at lowering the labour output ratio (labour saving technical progress).

A related question centres around the eponymous livelihood-lives trade-off. Lockdowns are a response to heightened infection rates. They shut-down components of the economy where the physical presence of the worker is required for production or involve the delivery of (mainly personal) services. In Eq. (5.1), we have made aggregate supply on the left-hand side of the equation a negative function of the infection rate. This is because the pandemic, left to its own devices, will potentially wreak havoc on the working population who complement the capital stock and intermediate goods needed in production, besides further attenuating aggregate demand for goods. The pandemic will inevitably engender a recession. For it to become a truly aggravated or *dark* recession,[6] associated with long lasting negative aggregate supply consequences, a *laissez faire* (do nothing) health response to the pandemic needs to be followed. In many countries, governments, at first, were hesitant about lockdown due to their ideological orientation and/or beliefs. This can worsen the negative economic impact of the pandemic by raising the probability and duration of recessions. Indeed, Deaton (2021) provides us with evidence that projected economic growth rates are negatively correlated with pandemic mortality rates. In other words, the economic down-turn due to the pandemic is worse in countries with greater COVID-19 related deaths.

Gans (2020) presents us with a pandemic possibility frontier (PPF) drawn in economic activity and health space; see Gans (2020) chapter 1. The PPF is concave to the origin and represents diminishing trade-offs between greater public health and economic activity at any point in time (Fig. 5.1). The pandemic shifts the PPF inwards, and if no public health response, such as lock downs are implemented, a dark recession with deeper PPF shrinkages emerge. Moreover, non-convexities or

---

[6] See Gans (2020) on dark recessions.

Fig. 5.1 Pandemic Possibility Frontier. (Source: Gans, 2020, Chapter 1)

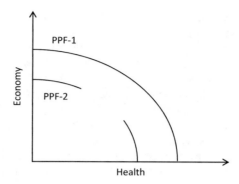

discontinuities emerge in the post-pandemic PPF, indicating unobtainable health-economy combinations, due to the dark recession. The recovery of the economy depends upon the avoidance of a dark recession, early action such as swift lockdowns, followed by targeted smart interventions such as test and trace, followed by mass vaccinations. This will allow the PPF to shift back to its original position. In the immediate aftermath of the pandemic, or its resurgence, any *lives-livelihood trade-off* is chiefly chimerical.

Dabla-Norris et al. (2015) and Ostry et al. (2014) show that the recent growth experiences of a cross-section of developed and developing countries suggest that excessive inequality is harmful to growth prospects. Hence, the belief that growth was assisted by inequality as a reward for risk and investment is much less valid in our time. In short, the *equality-growth trade-off* is weaker than before. This could be because greater inequality leaves economies more prone to financial crises, greater inequality results in less human capital accumulation, and because inequality contains within it the seeds of conflict, which is harmful for growth. In empirical models, redistributive policies, including social expenditures, appear to no longer harm growth prospects (Ostry et al., 2014) by crowding out private investment. Traditionally, it was believed that redistributive policies financed by taxation were distortion-inducing, even when they resulted in greater equity and social justice. Economic efficiency and equity needed to be *separated*, and furthermore there was an *efficiency–equity trade-off* (see Okun, 1975), which no longer seems to hold true.

The current pandemic will exacerbate poverty, particularly in developing countries. But the present work focusses on inequality, and it is still too early to work out the precise impact on metrics of income inequality, such as the GINI coefficient, because post-pandemic household surveys are yet to occur. Despite that, Furceri et al. (2020) have carried out hypothetical simulations based upon the measured effect of pandemics in the recent past, beginning with the SARS epidemic of 2003 to the Zika outbreak of 2016. They find that pandemics exacerbate income inequality in the 5 years that follow, and inequality measured in terms of net GINIs (after taxes and transfers) is even worse than market GINIs. This suggests that government interventions through the fiscal system have a regressive effect on the distribution of income. The channels that engender this inegalitarian outcome are not only worsening incomes of the poor, but also bleaker employment prospects for the less skilled, who are presumably pushed into more precarious employment.

The PEW Institute (2021) provides us with evidence of a shrinking global middle class, whom they define to be households where each member lives on between $10 and $20 a day in 2011 purchasing power parity (PPP)[7] dollars, using World Bank data and estimates on changes in income globally. They estimate that the global middle class will shrink by about 54 million individuals, with the bulk of these occurrences, about 60%, taking place in India. A declining middle class provides *prima facie* evidence of growing inequality and greater polarisation. Thus, we can say more about international inequality changes due to the pandemic compared to changes in national income distributions at this juncture, and this is what I turn to next.

5.4 Global Inequality

The interested reader can find a detailed elucidation of the concepts and measurement of global inequality in Milanovic (2016). A lot hinges on the unit of analysis, and whether it is the country or the household. In the former case, we take the average or per-capita income of a country as the metric for comparison. With the latter metric, the planet is treated as one country with households as the unit of analysis. If the country is the unit of analysis, measurement issues are more resolvable, as we have readily reliable data on per-capita incomes, whereas with the household method we need internationally comparable household surveys, which are not available prior to 1988.[8] Even with the country level analysis, which effectively treats everyone in the country as having its average income, there are two different methods. The first method, known as concept 1, is the inequality in average incomes across countries treating every country irrespective of population size as one equal unit, hence China and Antigua have the same weight. If we do apply population weights then we have concept 2 global inequality, and much of changes in this metric are driven by what happens in the more populous states of the world, such as China and India. Both these inequality measures decline when differences in cross-country average income decline, we have closing per-capita gaps between rich and poor nations.

Concept 1 international inequality (all countries given identical weights) kept creeping up until approximately 2000, when it started to decline because growth

---

[7] Purchasing power parity controls for differences in the cost of living internationally, compared to the United States. Thus, poorer countries, where the cost of living is lower get assigned a greater income compared to market exchange rates which tend to devalue living standards in poorer countries.

[8] With per-capita incomes a lot hinges on the PPP exchange rate year employed; the methodology of household surveys also raises issues because in some countries (developing Asia minus China, Africa) household income is made equivalent to household expenditure.

rates had recovered in all developing countries (compared to the 1980s and 1990s sub-Saharan African and Latin American nations fared poorly in their growth rates). For the first decade after 2000, growth rates in nearly all developing countries went ahead of the planet's the richer nations. Concept 2, population weighted international inequality has been steadily declining ever since China's growth resurgence in the 1980s, belatedly joined by India in the 1990s, as they collectively account for about 40% of the world's population.

Next, we have the concept of 'global' inequality, based upon the inequalities in household income across the planet. Global inequality may have declined by about 2 percentage GINI points between 1988 and 2008 to around 70.5 (Milanovic, 2016), but this finding, as the author points out, may mask the serious underestimation of the income of the top decile in the income distribution, who are often missed out in household surveys. In recent years the greatest beneficiaries of changes in the global income distribution have been the world's super rich (the top 1% in the income distribution), along with the middle classes in emerging market economies, chiefly in China and India; the greatest losers have been the lower middle and low income groups in developed countries, the traditional working class blue collar household, and more recently the hollowing out of the middle class in some rich countries.

Global inequality can be further decomposed into inequality within nation states, as well as the inequality between countries based on their average incomes. The inequality within nations has been rising everywhere, despite a short brake on inequality during the tenure of progressive regimes in Latin America between 2000 and 2010. The other component of global inequality is the inequality between nations, and here we can focus on the population weighted differences in per-capita income between nations. We would expect global inequality to fall as average income in the most populous developing countries (China and India) catches up with rich countries, and indeed this has been the case. But when we combine the two factors that make up global inequality, most of present-day global inequality still continues to be attributable to between country inequality, that is the inequality which stems from differences in average income between rich and poor nations. How rich or poor an individual depends on their position in the hierarchy of occupational incomes, but is more attributable to where they reside and work. A bus-driver may be poor because of his disadvantaged place in the wage league table, but more importantly it depends on where he is a bus-driver (a rich or poor country). Thus, according to Milanovic (2016) there is a *citizenship rent*, and consequently we may describe it as the *global inequality of opportunity*. Across the world, Milanovic (2016, figure 1.8) demonstrates that the share of billionaires wealth relative to global GDP was under 3% in 1987; this had increased to more than 6% by 2013. Accompanying this, the national income share of the middle class (defined as having an income in the range of 25% above and 25% below median national income) declined over this time period in nearly all Western democracies, with the United States exhibiting the lowest middle-class share, and the UK not far behind with the fourth lowest share (Milanovic, 2016, figure 4.8).

Global inequality tends to rise as globalisation accelerates (1870-1914 and after 1980), and declines when the international economy either de-globalises (as between 1914-45), or is managed in the style of the Bretton Woods era of 1945 to 1973. How will the current pandemic affect global inequality?

We do not have internationally comparable post-pandemic household surveys yet, but we have forecasts for growth in per-capita income, as well as COVID-19 induced national income changes for 2020. This allowed Deaton (2021) to investigate the impact of the pandemic on concept 1 and 2 global inequalities. First, Deaton notes that there were more reported COVID-19 fatalities in richer countries, and secondly GDP growth forecasts for 2021 by the international financial institutions is negatively related to COVID-19 mortality. China, both in terms of the reduction in its GDP, as well as COVID-19 deaths did better than most countries; whereas, India fared badly in both respects. Taken together, all of these stylised facts point to a fall in concept 1 (population unweighted) global inequality because the poorer economies of the world appear to have suffered less national income compression. Concept 2 (population weighted) global inequality, however, increases slightly. This, Deaton (2021) argues, is because China came out better than other countries, and as an upper-middle income country, increases in its relative per-capita income pushes up concept 2 global inequality. Given that within country income inequalities are set to rise as well, we may safely conjecture that global inequality has risen.

5.5 Conclusions: Inequality and Democratic Governance

The immediate impact of the COVID-19 pandemic is, therefore, likely to accelerate various forms of economic inequality in wealth and income. This is because the income of the poor was adversely affected more, both because of the already present trends in unskilled labour substitution, but also because the types of employment that the world's poor engage in was most severely disrupted by COVID-19, and the subsequent public health response. Certain countervailing income and job protection schemes can help, but it is mainly a short-term palliative. Population weighted international inequality has also risen because China's economy suffered less, and recovered faster, from the pandemic. In the short to medium-term there appears to be no livelihood-lives trade-off as countries with more COVID-19 mortality suffered greater national income contractions. Recent empirical evidence suggests that excessive inequality does not foster economic growth, and furthermore increasing social protection does not crowd out private investment, so the efficiency-equity trade-off is no currently inapplicable.

In the final analysis, the growth in inequality in the last four decades poses a menacing challenge to liberal-democratic governance. There is a rising wave of autocracy (VDEM report, 2021), especially in developing countries, accompanied by a rise in 'populism' in the context of democratic electoral processes, chiefly, but not exclusively, in developed countries. All of this occurring in the background of a highly globalised international economy, which helps to alter the domestic social contract in favour of the rich. The forces behind highly internationally mobile capital in a globalised context forces promote plutocratic policies and declining social protection in order to maintain international 'competitiveness'. Greater, and rising, inequality is the product of accelerating globalisation, and trade favours the highly skilled and owners of footloose capital. In the traditional Western democracies where the traditional working class, and in some instances even the lower-middle income groups have become pauperised and left bereft of hope there is a political backlash resulting in a vacuum which seemingly only populists can fill. Rodrik (2018) suggests that the rise in populism coincides with hyper-globalization (Murshed, 2020). The vote share of populist parties since 2000 in selected European and Latin American nations has exceeded 10% (Rodrik, 2018, figure 1). The crucial mechanism in the middle which helps transform increased globalisation into populist political success is inequality. Examples of recent populist victories include in the United States in 2016, in the UK in 2019 (along with Brexit referendum in 2016), Brazil in 2018 and in India in 2019. These populists are elected with nationalistic agendas, but do little to lessen inequality, while promoting illiberal and intolerant tendencies. This is the curious admixture of populism and plutocracy (Pierson, 2017).

In recent times, liberalism and democracy have become strange bedfellows. The majority can always tyrannize the minority in purely elective democracies unless constrained by liberal institutions and precepts. Excessive inequality, not only contains within it the seeds of conflict, it can also undermine the liberal aspects of democracy in settings where the electoral engine of democracy still appears to function as smoothly as ever by ushering in populist plutocrats into power.

References

Barro, R. J., Ursúa, J. F., & Weng, J. (2020). *The coronavirus and the great influenza pandemic: Lessons from the "Spanish Flu" for the coronavirus's potential effects on mortality and economic activity*. National Bureau of Economic Research (NBER) Working Paper 26866.

Dabla-Norris, E., Kochkar, K., Suphaphiphat, N., Ricka, F., & Tsounta, E. (2015). *Causes and consequences of income inequality: A global perspective*. IMF Staff Discussion Note.

Deaton, A. (2021). *COVID- 19 and global inequality*. National Bureau of Economic Research (NBER) Working Paper 28392.

Demena, B.A., Floridi, A., & Wagner, N. (2021). The short-term impact of COVID-19 on labour market outcomes: Comparative systematic evidence. In E. Papyrakis (Ed.), *Covid-19 and international development*. Springer.

Fantu, B., Demena, B.A., Genet, H., Lassooy, T. Y., Sathi, S., & Shuka, Z. S. (2021). Experiences of Eritrean and Ethiopian migrants during COVID-19 in the Netherlands. In E.Papyrakis (Ed.), *Covid-19 and international development*. Springer.

Furceri, D., Loungani, P., & Ostry, J. (2017). *The aggregate and distributional effects of financial globalization*. Mimeo, International Monetary Fund.

Furceri, D., Loungani, P., Ostry, J., & Pizzuto, P. (2020). Will Covid-19 affect inequality? Evidence from past pandemics. *Covid Economics, 12*, 138–157.

Gans, J. (2020). *Economics in the age of COVID-19*. MIT Press.

Hobsbawm, E. J. (1954). The general crisis of the European economy in the 17th century. *Past and Present, 5*(May), 33–53.

Kaplan, G., Moll, B., & Violante, G. L. (2020). *The great lockdown and the big stimulus: Tracing the pandemic possibility frontier for the U.S.* National Bureau of Economic Research (NBER) Working Paper 27794.

Milanovic, B. (2016). *Global inequality: A new approach for the age of globalization.* Belknap Press for Harvard University Press.

Murshed, S. M. (2020). *Populist politics and pandemics: Some simple analytics.* ISS working paper no. 664, https://repub.eur.nl/pub/131100/

Okun, A. (1975). *Equality and efficiency: The big trade-off.* Brookings Press.

Ostry, J., Berg, A., & Tsangarides, C. (2014). *Redistribution, inequality and growth.* IMF Staff Discussion Note.

PEW Research Center. (2021). *The pandemic stalls growth in the global middle class, pushes poverty up sharply*, 18 March 2021, https://www.pewresearch.org/global/2021/03/18/the-pandemic-stalls-growth-in-the-global-middle-class-pushes-poverty-up-sharply/. Accessed 23rd Mar 2021.

Pierson, P. (2017). American hybrid: Donald Trump and the strange merger of populism and plutocracy. *British Journal of Sociology, 68*(S1), S105–S119.

Piketty, T. (2014). *Capital in the Twenty-first Century.* Harvard University Press.

Polyani, K. (1944). *The great transformation.* Farrar and Rinehart.

Rodrik, D. (2018). Populism and the economics of globalization. *Journal of International Business Policy.* https://doi.org/10.1057/s42214-018-0001-4

Roemer, J. E. (1998). *Equality of opportunity.* Harvard University Press.

Romanello, M. (2021). Covid-19 and the informal sector. In E. Papyrakis (Ed.), *Covid-19 and international development.* Springer.

Routt, D. (2008). *The economic impact of the black death.* https://eh.net/encyclopedia/the-economic-impact-of-the-black-death/. Accessed 8th May 2020.

Scheidel, W. (2017). *The great Leveller: Violence and the history of inequality from the stone age to the twenty-first century.* Princeton University Press.

Stiglitz, J. (2012). *The Price of inequality: How Today's divided society endangers our future.* Norton.

Van Bergeijk, P. A. G. (2019). *Deglobalization 2.0: Trade and openness during the great recession and the great recession.* Edward Elgar.

VDEM Report. (2021). Alizada, N, Cole, R, Gastaldi, L, Grahn, S, Hellmeir, S, Kolvani, S, Lachapelle, J, Lührmann, A, Maerz, S.F., Pillai, S and Lindberg, S.I. 2021. Autocratization Turns Viral. Democracy Report 2021. University of Gothenburg: V-Dem Institute, https://www.v-dem.net/en/publications/democracy-reports/. Accessed 26th Mar 2021.

Voightländer, N., & Voth, H.-J. (2013). Gifts of Mars: Warfare and Europe's early rise to riches. *Journal of Economic Perspectives, 27*(4), 165–186.

Wagner, N. (2021). Indirect health effects due to COVID-19: An exploration of potential economic costs for developing countries. In E. Papyrakis (Ed.), *Covid-19 and international development.* Springer.

Chapter 6
The Short-Term Impact of COVID-19 on Labour Market Outcomes: Comparative Systematic Evidence

Binyam Afewerk Demena, Andrea Floridi, and Natascha Wagner

Abstract This chapter uses advanced meta-analysis techniques to evaluate the short-run impact of COVID-19 on various labour market indicators. Using 2429 reported estimates of labour market outcomes associated with COVID-19 from 29 empirical studies conducted in 12 countries, we show that large parts of the documented literature exhibit substantial publication bias. After controlling for publication bias, we find almost no practically meaningful impact of COVID-19 on earnings, hours worked and (un)employment. Next, we investigate if the identified publication bias is caused by publication characteristics. However, these characteristics do not appear to drive the identified bias. We also uncover the differences between developed and developing countries. The findings indicate that in practical terms both groups of countries experienced hardly any measurable short-term impacts of COVID-19 on the labour market. Furthermore, we detect systematic upward publication bias for unemployment and job loss in developed countries, and downward bias for employment, hours worked and earnings in developing countries. Overall, we concluded that in the short-run no major labour market effects could be identified related to formal employment.

6.1 Introduction

Disasters like epidemics or pandemics occur unexpectedly and can cause large shocks across the world not only to the health of those concerned but also to the organisation of life and daily activities including labour market dynamics. The onset of the global coronavirus pandemic (COVID-19) has resulted in a dramatic reduction in economic activity due to government introduced restrictions and individual behavioural reactions in the form of reduced mobility and economic activity because

---

B. A. Demena (✉) · A. Floridi · N. Wagner
Department of Development Economics, International Institute of Social Studies, Erasmus University Rotterdam, The Hague, The Netherlands
e-mail: demena@iss.nl

© The Author(s), under exclusive license to Springer Nature Switzerland AG 2022
E. Papyrakis (ed.), *COVID-19 and International Development*,
https://doi.org/10.1007/978-3-030-82339-9_6

of the fear of health risks (Crossley et al., 2020). Government induced lockdowns and individual behaviour adaptations disrupted the production processes of firms, and affected the demand for labour by the firms and the workers' ability and willingness to work. Therefore, COVID-19 has the potential of negatively impacting labour market outcomes by suddenly reducing both employers' labour demand and workers' labour supply (Kim et al., 2020).

Existing studies identify that health shocks from a pandemic impact labour market outcomes through direct and indirect channels (e.g., see Béland et al., 2020a). One direct channel stems from sickness and mortality as well as from care duties that deter workers, temporarily or permanently, from participating in the labour force. This direct channel corresponds to the destruction of human capital. Another direct channel is the increased risk and uncertainty about the future economy which may change investment behaviour. Consequently, employers may want to reduce labour costs and capital may be directed to essential industries (e.g., medical equipment manufacturers) or to areas that are relatively less affected by COVID-19 (Béland et al., 2020b). The indirect channel operates via changes in behaviour driven by the fear of infection, which leads to a fear of association with others and reduces the labour force participation. For instance, due to the fear of contagion and increased housework such as childcare and elderly care, workers may want to reduce working hours or drop out from employment (Kim et al., 2020).

These direct and indirect channels that trigger changes in labour demand and supply are likely to affect employment per se as well as working hours and wages; yet, the effects are a priori ambiguous since many channels are at work (Béland et al., 2020a, b; Kim et al., 2020). There is an emerging, and rapidly growing body of the empirical literature that has started exploring the effects of COVID-19 on labour market outcomes. However, the existing empirical studies have only reinforced the theoretical ambiguity, as they present contrasting findings which are far from being conclusive. We have reviewed the large number of published and unpublished empirical studies conducted until February 2021 and systematically identified 2429 reported estimates of labour market outcomes associated with COVID-19 from 29 empirical studies. A bird's-eye view of the primary empirical studies highlights the extent of disagreement in terms of the direction of impact and the statistical significance (Fig. 6.1). Approximately 58.2% find significant labour market impacts, whereas the other 40.8% report positive and negative, but insignificant impacts.

Given the divergent impact in terms of direction and significance, this chapter aims to contribute to the very recent debate by synthesizing the available empirical studies employing advanced meta-analysis techniques. In line with recent meta-studies in development economics (e.g., see Demena, 2015; Demena & Bergeijk, 2017; Bergeijk et al., 2019; Demena & Afesorgbor, 2020; Floridi et al., 2020, Floridi et al., 2021) the chapter will rigorously compile all available evidence by first assessing the overall average effect and significance of COVID-19 on labour market outcomes. Second, the chapter investigates whether the literature suffers from publication bias, and if so the extent of the bias. Next, we provide the genuine underlying impact after filtering out publication bias and last, we classify the empirical

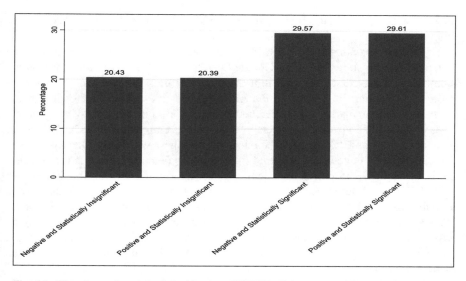

Fig. 6.1 Direction and statistical significance of COVID-19 impacts on labour market outcomes (N = 2429)

studies into developing and developed country analyses to provide a comparative perspective.

The remainder of the chapter is organised as follows: Sect. 6.2 describes the data collection and empirical approach in the context of the recent meta-analysis guidelines in Economics (Havránek et al., 2020). Section 6.3 details the results and Sect. 6.4 presents the comparative analysis between developed and developing countries. The final Section provides concluding remarks.

6.2 Data and Empirical Approach

6.2.1 Database Construction

For searching, coding, and analysing the reported and available empirical studies, we adopted the recently updated reporting guidelines for meta-analysis in economics (Havránek et al., 2020) and combine this guideline with the population-intervention-comparison-outcome-context (PICOC) protocol provided by Petticrew and Roberts (2008). The PICOC protocol was used as it fits well for policy-oriented systematic reviews (Floridi et al., 2020). Our *population* consists of the labour or workforce of a given country; the considered *impact or explanatory variable* is the incidence of the Coronavirus pandemic (COVID-19); we *compare* the condition of the labour force between the pre- and post-pandemic period and study as *outcomes* changes in employment levels, working hours, and income or earnings; the *context* is worldwide with a focus on comparative evidence between developed and developing countries.

To identify existing and accessible empirical studies, the initial stage of the extensive search started with the Google Scholar web engine. To reduce the large number of retrieved studies, we limit the search to those articles using the same words of the query in the text (*allintext*) or in the title (*allintitle*); for instance: "*allintext*: labour market". We searched using broad combinations of keywords: "COVID-19 + labour market effects", "pandemic + employment", "COVID-19 + hours of work", and "COVID-19 + job loss". Using these wide sets of keywords, we have identified a huge research output. Although this field of research is very recent as it deals with the current pandemic, for instance, the use of the keywords "COVID-19 + labour market effects", has led to about 19,000 potential studies to be initially inspected based on titles, abstracts, and keywords. Using Google Scholar, we also conducted a forward search to investigate studies that are already citing a given publication. Furthermore, hand searching was added to the internet search to include references that were found in the studies selected in the initial Google Scholar search.

The initial inspection of titles and keywords combined with a screening of abstracts in case of uncertainty resulted in 102 potential primary studies.[1] In general, the identified studies make use of variants of the following model to estimate the impact of COVID-19 on various labour market outcomes:

$$Y_{ipt} = \alpha + \beta PostCOVID_t + \mu X_{ipt} + \theta_{ipt} + \varepsilon_{ipt} \tag{6.1}$$

where Y is a measure of a labour market outcome of individual i in region or province p at time t. X represents the vector of observable characteristics to control for individual characteristics, such as gender, age, and the level of education, θ controls for unobservable fixed effects, such as provincial fixed effects and time fixed effects and ε is the error term. The onset of the COVID-19 pandemic is captured by the variable *PostCOVID* that takes the value of 1 for all months following the local outbreak of COVID-19 (in most cases all months after February 2020) and zero otherwise. In such a model, our variable of interest is the parameter β, representing the average labour market response to the pandemic.

Most first-stage studies from title and abstract screening were irrelevant as we used an inclusive search query that resulted in a large set of false hits, and hence we dropped 44 studies after examining the introductions and conclusions of the initially identified 102 studies. After having identified the potential primary studies, in the second stage, the review proceeded with the selection of articles satisfying the inclusion criteria for a detailed review. We strictly apply the following inclusion criteria: (i) English language and (ii) empirical micro-econometric studies that report (iii) regression-based coefficients of the *PostCOVID* variable on a given labour market outcome along with a standard error or t-*value* as well as the sample

---

[1] We employed a similar procedure as specified by Demena (2017) and Floridi et al. (2020).

size. In this second stage screening of the identified studies, we conduct an in-depth examination of the full texts. This process led to the inclusion of 19 eligible studies. In addition the hand-search of the reference lists of these papers led to 10 additional primary studies resulting in a dataset of 29 studies for 12 countries[2] published until February 2021. Figure 6.2 provides the flowchart (PRISMA) of the 3-stage study selection process: Identification, Screening and Eligibility.

After establishing study eligibility a Microsoft Excel template was designed for coding. Coding was conducted on several characteristics related to measures and effects of COVID-19, as well as study and journal publication quality features. In line with meta regression analysis (MRA) requirements, the selection of the primary studies was conducted by two researchers independently to reduce bias due to human error. Coding was initially done by AF and then independently reviewed by BD to check the consistency. After double-checking the initial data and consensus on disagreements, the dataset was transferred to a Stata file for analysis.

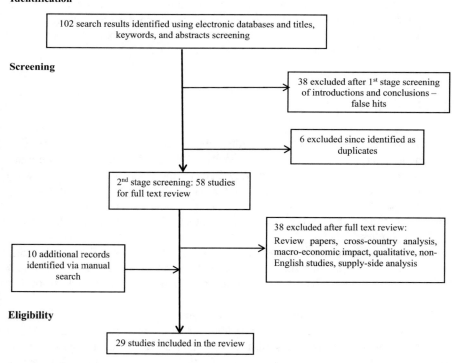

Fig. 6.2 Schematic flowchart of the empirical studies' selection process (PRISMA). (Source: Adapted from Demena et al., 2020)

---

[2] These are Australia, Bangladesh, Canada, Colombia, Greece, India, Mexico, the Netherlands, Singapore, Spain, the United Stata of America, and Vietnam.

6.2.2 Putting the Collected Estimates Together

We have collected various labour outcome indicators to systematically assess the impact of the recent pandemic. Because of the various proxies for labour market outcomes, it is necessary to use some standardisation approach to make the reported COVID-19 estimates comparable across primary studies. Following recent meta-analyses (e.g., Floridi et al., 2020; Floridi et al., 2021), we use the partial correlation coefficient (PCC) to make the estimated effect of COVID-19 on labour market outcomes comparable across studies. We compute the PCC as follows:

$$PCC_{is} = \frac{t_{is}}{\sqrt{t^2_{is} + df_{is}}} \quad (6.2)$$

where PCC_{is} represents the partial correlation coefficient between the COVID-19 indicator and any given measure of a labour market outcome with i denoting the reported regression specification of the primary study s; t_{is} is the associated t-value of the reported coefficient and df are the degrees of freedom of each estimate's regression. Using the PCCs, we are able to provide summary statistics across the different reported indicators of labour market outcomes.

6.2.3 Dataset

The analysis exploits a meta-dataset with 2429 estimates from 29 studies. The average and median number of estimates per study are 82 and 70, whereas the minimum and maximum are 2 and 415, respectively. Because the pandemic spread worldwide during the first quarter of 2020, the literature is very recent: the oldest study dates back to April 2020, and the most recent one is from January 2021. Relatedly, due to the recent flourishing of the empirical literature we identified only four peer-reviewed studies.

The analysed empirical literature is mainly geared toward developed countries: we retrieved 24 studies focusing on developed countries (1901 observations) which are considerably more than those on developing countries (463 effects from five studies).

The included studies focus on a variety of labour market indicators; we included in our analysis five main outcomes: employment, hours worked, earnings, job loss and unemployment. Roughly one of three estimates derives from employment indicators (33.8%), followed by hours worked (32.7%), and job loss (14.7%). Earnings make up for 10.4% and unemployment for 8.4% of the estimates. We then sort the five outcome indicators in two groups based on the expected signs of change. On the one hand employment, hours worked, and earnings are expected to decrease due to the outbreak of Covid-19 and hence display negative coefficients; on the other hand, unemployment and job losses are expected to be positively affected as they are likely to increase due to the pandemic.

6.2.4 Empirical Strategy

Funnel Plots

The reported estimates might be subject to publication selection bias, which is known as common tendency in empirical research (Stanley & Doucouliagos, 2012). Publication bias arises when researchers, journal editors and referees prefer or expect specific reported signs of the estimates or just higher statistical significance regardless of their sign. In the field of economics, various meta-analysts have already reported strong publication bias. For instance, Demena (2015) and Demena and Bergeijk (2017) identified positive bias in the literature of foreign direct investment and spillovers; Bergeijk et al. (2019) found bias on the economic determinants of sanctions and Floridi et al. (2020) showed positive publication bias in the literature about the formalisation of informal firms.

To detect whether the primary studies considered here are affected by choice-driven biases, we start with a graphical inspection of the estimates reported. We use a common graphical tool: the funnel plot is a scatterplot that presents the inverse of the standard error on the y-axis and the effect size on the x-axis and looks like an inverted funnel (Demena, 2014). However, the funnel plot is subjective since it can only be visually inspected, and it cannot be used to conclusively determine the absence or presence of publication bias which requires a more formal statistical approach.

Statistical Analysis

To detect and correct artifacts of choice-driven biases we employ a more powerful econometric approach, namely the meta-regression model (MRM). More specifically, the following equation is estimated:

$$PCC_{is} = \beta_0 + \beta_1 SE_{pccis} + e_{is} \tag{6.3}$$

where PCC_{is} is the measure of the labour market outcome derived using Eq. (6.2) and SE_{pccis} is the associated standard error, e_{is} is the error term of the mete-regression. The parameter β_0 is the genuine underlying effect and β_1 denotes publication bias. This basic MRM shows that as sample size increases and thus the available quantity of information increases, SE_{pccis} will approach zero (Demena, 2014). That implies that with large samples the PCC_{is} will approach the intercept in Eq. (6.3), implying that the underlying genuine effect is corrected for publication bias. Put differently, in the absence of bias, the overall genuine effect should vary randomly around the intercept β_0 and should be independent of the standard error SE_{pccis} (Stanley & Doucouliagos, 2012). Conversely, the presence of selection bias can be detected if reported estimates of labour market outcomes are correlated with their standard errors (Demena, 2017).

As already introduced in the data section, the reported estimates of labour market outcomes are derived from various econometric designs, specifications, and sample sizes. Hence Eq. (6.3) can potentially suffer from heteroskedasticity implying that plain vanilla OLS should not be applied. To reduce this problem, Stanley and Doucouliagos (2012) suggest that Eq. (6.3) should be estimated with a weighted least squares (WLS) approach using as weights the inverse of the variance of the estimated PCC_{is}. This results in the following estimating equation:

$$t_{is} = \beta_1 + \beta_0 \left(1 / SE_{pccis}\right) + u_{is} \tag{6.4}$$

where t_{is} denotes the t-statistic measuring the statistical significance of the PCC. Testing the null hypothesis that $\beta_1 = 0$ allows us to investigate the issue of publication bias, which is known as the funnel asymmetry test (FAT). Rejecting the null hypothesis indicates the presence of selection bias. Similarly, testing the null hypothesis that the slope in Eq. (6.4) is zero, i.e., $\beta_0 = 0$, we estimate the genuine underlying effect (if any) after filtering out publication bias, which is known as precision effect test (PET). Rejecting the null hypothesis implies that there is a genuine underlying effect.

Econometric Concerns

A major concern in estimating Eq. (6.4) is the issue of within-study dependence when multiple estimates are collected from each study. We have gathered multiple estimates for a given study which are unlikely to be statistically independent (Demena, 2015). To correct for within-study dependence, we first employ OLS combined with clustered standard errors (CDA) after the transformation into the WLS model. Next, we use a fixed effect (FE) estimation that can also address within-study variation. Beyond within-study dependence, however, between-study dependence is also an important concern because multiple primary studies are published by the same authors (and thus estimates derived from two or more studies published by the same authors are unlikely to be statistically uncorrelated). To account for this, we employ the multi-effect, multi-level model (MEM) since contrary to the OLS and FE models it accounts for both dependencies jointly. Thus, we present the CDA approach as baseline result and augment it in a next step with FE. Then we conduct the MEM; in interpreting our results we give most weight to the latter model.

Another potential empirical concern is related to outliers in the reported estimates that may distort or drive the results. We applied the multivariate outlier method proposed by Hadi (1994). We identified about 2.67% of the observations as outliers.

6.3 Results and Discussion

6.3.1 Overall Average Effect

We start the discussion of the results with the introduction of the summary statistics. Table 6.1 reports the full sample summary statistics of the impact of COVID-19 on the various labour market outcomes taken together. The simple average effect is −0.003 with a 95% confidence interval of [−0.004; −0.001], implying a negative and statistically significant impact. However, applying the meta-analysis guidelines by Doucouliagos (2011), the overall impact has limited practical significance.[3] The summary statistics using inverse variance weights suggest a similar picture. To refine the assessment, we narrow down our analysis to sets of more homogenous indicators or groups. The first group consists of outcome variables that report a change in earnings received, hours worked and employment. The second group consists of the categories job loss and change in unemployment. The overall results are consistent with the findings for the full sample, but the sub-sample analysis provides a more nuanced picture of the effects. The results associated with more homogenous groups of indicators is that the impact on earnings, hours worked, and employment is negative whereas on job loss and unemployment is positive. Overall, despite being statistically significant all estimates are practically meaningless.

Even though these averages provide insightful summaries, they are rather basic. As indicated in the empirical approach section, we need to account for publication bias to provide a purged overall genuine effect of COVID-19 on labour market outcomes.

Table 6.1 Average impact of formality on performance

Method	Effect size	S.E.	95% confidence interval	
Simple-average[a]				
All-estimates	−0.003	0.001	−0.004	−0.001
Earnings, hours worked and employment	−0.007	0.001	−0.009	−0.005
Job loss and unemployment	0.010	0.001	0.008	0.012
Weighted-average[b]				
All-estimates	0.0005	0.0003	−0.0002	0.001
Earnings, hours worked and employment	−0.002	0.0004	−0.002	−0.001
Job loss and unemployment	0.004	0.001	0.003	0.005

Note: [a]arithmetic mean of the PCC. [b]inverse variance as weight

---

[3] The guideline suggests that a MRA coefficient is small if it is at most 0.07 in absolute terms, it is of medium size if it ranges around 0.17 in absolute terms, and large if it is at least 0.33 in absolute terms.

6.3.2 Graphical Inspection

We start with the visual approach to publication bias making use of a funnel plot. The vertical axis displays the logarithm of the inverse of the standard error (precision) and the horizontal axis provides the estimated effect size of the impact of COVID-19 on the various labour market outcomes. In the absence of publication bias, the plot should be symmetrical with less precise estimates or larger standard errors estimated from small sample sizes being widely dispersed at the bottom of the graph (Demena, 2014). Whereas the most precise estimates or small standard errors estimated from large samples should be closely distributed around the underlying effect at the top of the graph. Moreover, both negative and positive estimates should be reported regardless of their statistical significance (Demena et al., 2021; Bergeijk et al., 2019). In contrast, in the presence of publication bias, the funnel plot is asymmetrical, implying some estimates are discarded and/or others are overrepresented (Demena, 2017).

Figures 6.3 and 6.4 show the funnel plots demonstrating the relationship between COVID-19 and labour market outcomes. Both plots seem to exhibit symmetrical distributions suggesting the absence of publication selection bias in the reported estimates (Figs. 6.3 and 6.4). Specifically, the sub-sample of earnings, hours worked, and employment is symmetrically distributed around their most precise estimate,

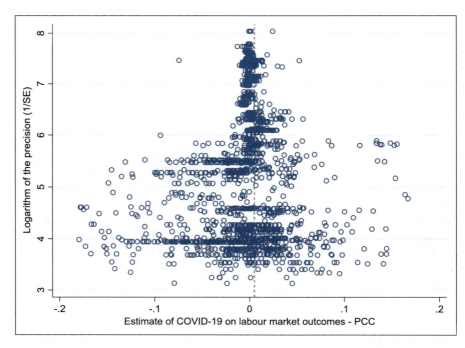

Fig. 6.3 Funnel plot of the impact of COVID-19 on labour market outcomes – Full sample
Note: The dashed vertical line represents the overall weighted full sample mean as reported in the lower panel of Table 6.1.

6 The Short-Term Impact of COVID-19 on Labour Market Outcomes: Comparative... 81

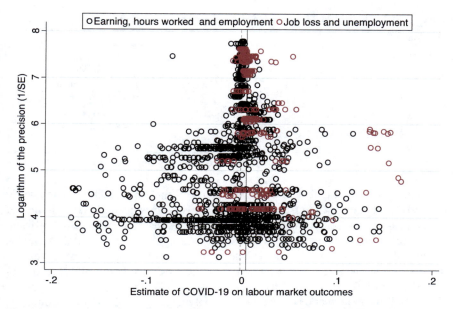

Fig. 6.4 Funnel plot of the impact of COVID-19 on two sub-groups of labour market outcomes
Note: The dashed and solid vertical lines represent the overall weighted mean for the sub-sample of earnings, hours worked, and employment (group 1) and the sub-sample of job loss and unemployment (group 2), respectively. The lines represent the averages that are reported in the lower panel of Table 6.1

which is −0.002 and represented by the dashed vertical line (Fig. 6.4). However, the sub-sample of job loss and unemployment outcomes seems to have a fluffy concentration of very few estimates on the right-hand-side compared to no such concentration of estimates on the left-hand-side of the funnel plot. That cluttering of estimates represents a much larger effect compared to the most precise estimates that are concentrated around the average of 0.004, which is represented by the solid vertical line. This suggests a slight preference towards positive estimates (Fig. 6.4). Yet, note that the interpretation of the funnel plot is subjective (visual inspection) and cannot be used to conclusively determine the absence or presence of publication bias. Therefore, we now apply a more formal statistical approach.

6.3.3 Meta-Regression Analysis

To investigate the presence of publication bias and provide the genuine underlying impact more formally, we perform the FAT and PET estimating Eq. (6.4) with three empirical specifications. As pointed out in the empirical strategy section, we present the CDA approach as baseline result and augment it in a next step with FE. Yet, in discussing the results we mainly resort to the most rigorous model with MEM. Table 6.2 reports the results associated with the full sample and Table 6.3 represents findings for the two sub-groups of labour market indicators.

Table 6.2 MRA for FAT-PET: Publication bias and genuine effect – Full sample

Full sample						
Variables	(1)		(2)		(3)	
	CDA		FE		MEM	
	Coefficient	S. E.	Coefficient	S. E.	Coefficient	S. E.
Bias (FAT)	0.017	0.001	1.591*	0.850	2.885**	1.282
Genuine effect (PET)	0.000	0.001	−0.002	0.001	−0.002***	0.000
Observations	2364		2364		2364	
Studies	29		29		29	

Note: ***, **, * stand for a 1%, 5%, and 10% level of statistical significance, respectively

First, we discuss the results of the full sample (Table 6.2). Across econometric methods (exception is CDA), the FAT consistently reports upward bias as indicated by the positive and statistically significant estimates that range between 1.591 and 2.885. According to Doucouliagos and Stanley (2013),[4] the size of publication bias found in this literature is substantial since it ranges between 1 and almost 3. Next we turn to the genuine effect that is purged of publication bias. In our preferred model (MEM), the PET test suggests the presence of a statistically significant impact of COVID-19 on labour market outcomes. Yet, applying the guidelines by Doucouliagos (2011), the overall impact has limited practical significance. This finding is consistent with the weighted average effect reported in Table 6.1. In short, the FAT test provides evidence for substantial upward bias, and the PET test suggests a weak practical impact.

To further refine the assessment, Table 6.3 presents results for the sub-samples with the more homogeneous groups of outcome indicators. These results provide a more nuanced picture of the labour market impact of COVID-19. In Panel 1 that presents estimates associated with the sub-group of earnings, hours worked and employment we find similar PET results as for the full sample, implying a negative and significant impact of limited practical meaning. Importantly, in this sub-sample analysis, we do not detect the presence of publication selection bias. In contrast, the analysis in Panel 2 that presents estimates associated with the sub-group of job loss and unemployment we show that there is substantial publication bias that is even bigger than the one reported in Table 6.2. This result is consistent with the visual findings from the funnel plot in Fig. 6.4. In other words, the researchers of the primary studies about job loss and unemployment were predisposed to select specifications that find adverse effects associated with the onset of the pandemic. The PET results for this second sub-group indicate statistically significant and positive impacts of COVID-19 on job loss and unemployment, yet in terms of magnitude they identify almost no practical significance.

In sum, our results point out that there are negative (positive) impacts of COVID-19 on earnings, hours worked and employment levels, (job loss and unemployment). Concomitantly, both sub-groups of indicators show only very small impacts with hardly any practical significance. It needs to be noted here that all

---

[4] Publication bias is 'little to modest' if FAT is smaller than 1, it is 'substantial' if it ranges between 1 and 2, and 'severe' if it is at least 2 (Doucouliagos & Stanley, 2013).

Table 6.3 MRA for FAT-PET: Publication bias and genuine effect – two sub-groups of labour market outcomes

Panel 1 – Earnings, hours worked and employment						
Variables	(1)		(2)		(3)	
	CDA		FE		MEM	
	Coefficient	S. E.	Coefficient	S. E.	Coefficient	S. E.
Bias (FAT)	−0.592	1.216	−0.194	0.491	−0.544	0.929
Genuine effect (PET)	−0.001	0.001	−0.001	0.001	−0.001***	0.000
Observations	1818		1818		1818	
Studies	21		21		21	
Panel 2 – Job loss and unemployment						
Variables	(1)		(2)		(3)	
	CDA		FE		MEM	
	Coefficient	S. E.	Coefficient	S. E.	Coefficient	S. E.
Bias (FAT)	4.451**	1.895	6.355***	1.846	8.602***	1.954
Genuine effect (PET)	0.000	0.002	0.002	0.002	0.002***	0.001
Observations	1274		1274		1274	
Studies	20		20		20	

Note: ***, **, * stand for a 1%, 5%, and 10% level of statistical significance, respectively

studies included in this analysis present short-term impacts of COVID-19 on the labour market. Thus, it is plausible that in the medium to long-run labour market effects will materialize. For instance, it has been stressed that the adverse effects of the pandemic on the GDP can become larger through a multiplier effect that is triggered by the economic slowdown (Kohlscheen et al., 2020). Spillovers and spillbacks across regions can magnify the size and length of an economic shock (Kohlscheen et al., 2020); thus, in the long-run even countries that have been less hit initially may experience negative economic consequences. Moreover, in the long-run pandemics may reduce the labour force and induce precautionary savings (Jordà et al., 2020). Lastly, medium to long-term effects of a pandemic may be determined by structural and institutional differences among countries (Fana et al., 2020) implying that some countries may require more time to recover from the crisis compared to others. In turn, Jordà et al. (2020) argues that pandemics do not shrink the real GDP in the long-run as labour productivity increases.

While the literature on the effects of pandemics is similarly inconclusive, we can address whether the identified publication bias is caused by publication characteristics of the primary studies or other moderator variables.

6.3.4 Does Publication Bias Arise Due to Publication Characteristics?

Publication bias can be driven by publication characteristics. For instance, the editors or referees of peer-reviewed journals might prefer larger reported estimates or researchers of the primary studies might pre-select large estimates for journal submissions. Similarly, if researchers have a preference for identifying large estimates, they might cite studies that deliver such estimates (Bajzik et al., 2020). To be specific, the idea is to test for the driving force of publication bias investigating whether the slope in Eq. (6.3) is larger for peer-reviewed, newly published, and highly cited studies. We agree that there is no bullet-proof to establish causality, however, we find that it is important to gauge what are the driving factors of publication bias.

If publication characteristics cause the reported upward bias, their interactions with the explanatory variable SE_{pccis} in Eq. (6.3) will be positive and significant (Demena & Bergeijk, 2017). To better visualize the results and keep the presentation of the results manageable, in Table 6.4, we report only results related to the publication characteristics and their interaction with the standard errors SE. The results suggest that merely one out of the three interactions (Column 4) is significant at the 5% level when jointly included in the regression specification. It is the interaction between the age of the publication and the standard error. Moreover, this interaction is associated with downward bias as compared to the reported upward bias. Importantly, the inclusion of the moderator variables does not remove the reported publication bias. The coefficient is still large and statistically significant (5.848, p-value<0.1). This finding indicates that our results of upward publication bias is not driven by the influence of publication characteristic but potentially by other sources of heterogeneity that are not considered in this study. Indeed, the heterogeneity across all the reported estimates is evident by Cochran's Q-test. The χ^2 distribution with $N\text{-}1$ degrees of freedom and 2363 reported estimates is 57,801 (p-value = 0.000). The I^2 test of heterogeneity reports that the variation in the reported labour market outcomes associated with COVID-19 is to 98.3% attributable to heterogeneity in the moderator variables. This is a significant opportunity for

Table 6.4 Potential mediating factors of publication bias

Variables	(1) Published (P)	(2) Pub. Age (Y)	(3) Citation (C)	(4) P + Y + C
Constant	−0.016**	−0.044**	0.008	−0.001
	(0.008)	(0.020)	(0.014)	(0.048)
SE	1.508*	5.986***	0.700	5.848*
	(0.853)	(1.424)	(1.011)	(3.233)
SE * Published	−1.784**			−0.311
	(0.907)			(1.747)
SE * Pub. Age		−1.083***		−0.876**
		(0.275)		(0.445)
SE * Citations			0.043	−0.189
			(0.025)	(0.710)
Observations	2364	2364	2364	2364
Studies	29	29	29	29

Notes: ***, **, * stand for a 1%, 5%, and 10% level of statistical significance, respectively. Standard errors are reported in parentheses. Reported results are estimated employing the specification reported in Eq. (6.3) with MEM inverse variance weights. The publication variables are also included directly but the results are not reported for the sake of brevity

future research to further explorer other potential sources of heterogeneity in driving the estimates reported by the primary studies.

6.4 Comparative Perspective of Developed Versus Developing Countries

The last step of our analysis consists in comparing the effects of the Covid-19 pandemic on labour markets in developed versus developing countries. We performed the FAT-PET for the two country sub-samples and reported the results by type of outcome indicator in Table 6.5. Even at a glance, the findings indicate once more the heterogeneity of the reported findings, as the impact of COVID-19 on labour markets varies depending on the considered sub-group of countries and labour indicator.

For the developed countries, estimates referring to earnings, hours worked and employment display a small negative genuine effect that is statistically significant. In addition, we identify a non-significant, downward publication bias. In turn, job loss and unemployment estimates report a small and non-significant negative

Table 6.5 MRA for FAT-PET: Publication bias and true effect – across groups of countries

Earnings, hours worked and employment – Developed countries						
Variables	(1)		(2)		(3)	
	CDA		FE		MEM	
	Coefficient	S. E.	Coefficient	S. E.	Coefficient	S. E.
Bias (FAT)	0.544	1.245	0.756	0.547	−0.178	1.117
Genuine effect (PET)	−0.001	0.001	−0.001	0.001	−0.001***	0.000
Observations	1371		1371		1371	
Studies	16		16		16	
Job loss and unemployment – Developed countries						
Variables	(1)		(2)		(3)	
	CDA		FE		MEM	
	Coefficient	S. E.	Coefficient	S. E.	Coefficient	S. E.
Bias (FAT)	4.417**	1.899	6.299***	1.842	8.670***	2.092
Genuine effect (PET)	−0.000	0.002	−0.002	0.002	−0.002	0.001
Observations	530		530		530	
Studies	14		14		14	
Earnings, hours worked and employment – Developing countries						
Variables	(1)		(2)		(3)	
	CDA		FE		MEM	
	Coefficient	S. E.	Coefficient	S. E.	Coefficient	S. E.
Bias (FAT)	−3.980	2.033	−5.678**	1.345	−3.366*	1.743
Genuine effect (PET)	0.002	0.002	0.006	0.003	0.004**	0.002
Observations	447		447		447	
Studies	5		5		5	

Note: ***, **, * stand for a 1%, 5%, and 10% level of statistical significance, respectively

genuine effect along with a large and statistically significant upward publication bias. For developing countries, we have considerably fewer estimates and therefore only present the joint analysis of the meta-effects for job loss, working hours and employment. We identify a small positive genuine effect and a large downward publication bias.

The findings indicate that studies about developed countries are on average more affected by publication bias compared to those from developing countries. Yet, no doubt, publication bias is an issue across studies. Moreover, across settings we identify hardly any practically meaningful effects of COVID-19 on the labour market. One possible explanation is that across settings the adverse effects of COVID-19 have yet to materialize. For instance, in the medium to long-run pandemics may lead to economic recession due to the contraction of foreign direct investments (Kohlscheen et al., 2020).

A short-coming of the comparison between developed and developing country labour markets is that the latter are characterised by a dual labour market with high levels of informality and informal employment representing the main form of labour (Alon et al., 2020 as well as Murshed, 2021 and Romanello, 2021). Our analysis focuses only on the effects of COVID-19 on the formal labour market, leaving out the impact on informal employment. Thus, it is plausible that the effects of the pandemic on informal employment are more pronounced as lockdown policies might hit informal workers harder (Alon et al., 2020) because they have limited access to social security and to subsidies (Khambule, 2020).

Another limitation of the study at hand is that we do not account for the intensity of the pandemic. It could have played a crucial role: countries where the pandemic was less intense or better managed may have perceived smaller adverse effects.

6.5 Concluding Remarks

We have reviewed the large number of published and unpublished empirical studies about the short-term impacts of COVID-19 on the labour market that were carried out until February 2021. This systematic review has identified 2429 reported estimates of short-term labour market outcomes associated with COVID-19 from 29 empirical studies conducted in 12 countries. A bird's-eye view of the primary empirical studies shows the plethora and often conflicting findings of this literature. This chapter, therefore, synthesised the available empirical studies with advanced meta-analysis techniques. To give a fair presentation of the literature, the MRA, on purpose included different dimensions of labour market outcomes reaching from earnings to hours worked and employment including as well job loss and unemployment.

We found a statistically significant negative impact of COVID-19 on earnings, hours worked and the employment level, and a positive impact on job loss and unemployment. Yet, for both groups of indicators we cannot establish any practical significance of the findings. It is plausible that a medium to longer time period is needed for labour market effects to materialize and the presented short-term impacts are too early (Fana et al., 2020; Jordà et al., 2020; Kohlscheen et al., 2020). In addition, we found substantial upward publication bias in particular for estimates reported for job loss and unemployment indicators. Exploring this aspect further, we show that the publication characteristics of the primary studies do not appear to be driving the identified publication bias.

We also studied whether there are differences across different groups of countries: developed versus developing countries. We found that estimates from developed countries report a small negative effect on employment, hours worked and earnings and display downward non-systematic publication bias. In turn, we detect small non-significant negative effects on unemployment and job loss, and large, statistically significant upward publication bias for developed countries. Estimates for the group of developing countries display small positive effects on employment, hours worked and earnings with a large downward publication bias. Overall, the results indicate that the empirical literature reports small short-term effects on labour market outcomes on average and that negative effects of Covid-19 are particularly evident for developed countries. This difference could be explained by the lower intensity of Covid-19 in developing countries; by demographic factors such as the younger average age of the populations of developing countries; and by economic and structural differences across the two groups of countries.

To conclude, the findings of our analysis are far from being conclusive, but rather represent a first step for future studies. More research is needed and expected that explores the long-run effects of the pandemic and there seems to be a need to watch out for desirability bias in publications.

References

Alon, T. M., Kim, M., Lagakos, D., & VanVuren, M. (2020). *How should policy responses to the covid-19 pandemic differ in the developing world?*, National Bureau of Economic Research (No. w27273).

Bajzik, J., Havranek, T., Irsova, Z., & Schwarz, J. (2020). Estimating the Armington elasticity: The importance of study design and publication bias. *Journal of International Economics, 127*, 103383.

Béland, L. P., Brodeur, A., & Wright, T. (2020a). *The short-term economic consequences of Covid-19: Exposure to disease, remote work and government response.* IZA Discussion Papers, No. 13151.

Béland, L. P., Brodeur, A., & Wright, T. (2020b). *COVID-19, stay-at-home orders and employment: Evidence from CPS data.* IZA Discussion Papers, No. 13282.

Crossley, T. F., Fisher, P., & Low, H. (2020). The heterogeneous and regressive consequences of COVID-19: Evidence from high quality panel data. *Journal of Public Economics, 193*, 104334.

Demena, B. A. (2014). *New wine in old bottles: A meta-analysis of FDI and productivity spillovers in developing countries.* Conference Proceeding on Economic, Social and Political Developments and Challenges in IGAD region (pp. 5–27). The Horn Economic and Social Policy Institute. https://www.hespi.org/2015/01/25/igadec-conference-proceeding-2015/.

Demena, B. A. (2015). Publication bias in FDI spillovers in developing countries: A meta-regression analysis. *Applied Economics Letters, 22*(14), 1170–1174.

Demena, B. A. (2017). *Essays on intra-industry spillovers from FDI in developing countries: A firm-level analysis with a focus on Sub-Saharan Africa*, PhD diss., Erasmus University, The Hague.

Demena, B. A., & Afesorgbor, S. K. (2020). The effect of FDI on environmental emissions: Evidence from a meta-analysis. *Energy Policy, 138*, 111192.

Demena, B. A., & van Bergeijk, P. A. G. (2017). A meta-analysis of FDI and productivity spillovers in developing countries. *Journal of Economic Surveys, 31*(2), 546–571.

Demena, B. A., Artavia-Mora, L., Ouedraogo, D., Thiombiano, B. A., & Wagner, N. (2020). A systematic review of mobile phone interventions (SMS/IVR/calls) to improve adherence and retention to antiretroviral treatment in low-and middle-income countries. *AIDS Patient Care and STDs, 34*(2), 59–71.

Demena, B. A., Reta, A., Jativa, G. B., Kimararungu, P., & van Bergeijk, P. A. (2021). *Publication bias of economic sanction research: A Meta-analysis of the impact of trade-linkage, duration and prior relations on sanction success*, Edward Elgar (forthcoming).

Doucouliagos, H. (2011). *How large is large? Preliminary and relative guidelines for interpreting partial correlations in economics* (No. 2011_5). Deakin University, Faculty of Business and Law, School of Accounting, Economics and Finance.

Doucouliagos, C., & Stanley, T. D. (2013). Are all economic facts greatly exaggerated? Theory competition and selectivity. *Journal of Economic Surveys, 27*(2), 316–339.

Fana, M., Pérez, S. T., & Fernández-Macías, E. (2020). Employment impact of Covid-19 crisis: From short term effects to long terms prospects. *Journal of Industrial and Business Economics, 47*(3), 391–410.

Floridi, A., Demena, B. A., & Wagner, N. (2020). Shedding light on the shadows of informality: A meta-analysis of formalization interventions targeted at informal firms. *Labour Economics, 67*, 101925.

Floridi, A., Demena, B. A., & Wagner, N. (2021). The bright side of formalization policies! Meta-analysis of the benefits of policy-induced versus self-induced formalization. *Applied Economics Letters.* https://doi.org/10.1080/13504851.2020.1870919.

Hadi, A. S. (1994). A modification of a method for the detection of outliers in multivariate samples. *Journal of the Royal Statistical Society: Series B (Methodological), 56*(2), 393–396.

Havránek, T., Stanley, T. D., Doucouliagos, H., Bom, P., Geyer-Klingeberg, J., Iwasaki, I., Reed, W. R., Rost, K., & Van Aert, R. C. M. (2020). Reporting guidelines for meta-analysis in economics. *Journal of Economic Surveys, 34*(3), 469–475.

Jordà, Ò., Singh, S. R., & Taylor, A. M. (2020). *Longer-run economic consequences of pandemics*, National Bureau Economic Research (No. w26934).

Khambule, I. (2020). The effects of COVID-19 on the south African informal economy: Limits and pitfalls of government's response. *Loyola Journal of Social Sciences, 34*(1), 95–109.

Kim, S., Koh, K., & Zhang, X. (2020). *Short-term impacts of COVID-19 on consumption and labor market outcomes: Evidence from Singapore*. IZA Discussion Papers, No. 13354.

Kohlscheen, E., Mojon, B., & Rees, D. (2020). *The macroeconomic spillover effects of the pandemic on the global economy*. Bank for International Settlements (No. 4).

Murshed, S. M. (2021). Consequences of the Covid-19 pandemic for economic inequality. In E. Papyrakis (Ed.), *Covid-19 and international development*. Springer.

Petticrew, M., & Roberts, H. (2008). *Systematic reviews in the social sciences: A practical guide*. John Wiley and Sons.

Romanello, M. (2021). Covid-19 and the informal sector. In E. Papyrakis (Ed.), *Covid-19 and international development*. Springer.

Stanley, T. D., & Doucouliagos, C. (2012). *Meta-regression analysis in economics and business*. Routledge.

van Bergeijk, P. A. G., Demena, B. A., Reta, A., Jativa, G. B., & Kimararungu, P. (2019). Could the literature on the economic determinants of sanctions be biased? *Peace economics. Peace Science and Public Policy, 25*(4), 636–650.

Chapter 7
Covid-19 and the Informal Sector

Michele Romanello

Abstract The most appropriate strategy to contain the spread of Covid-19 according to the WHO, government authorities and health experts is a set of measures that include social isolation, quarantine and lockdown, in addition to hygienic practices. Yet, developing countries have certain characteristics that make it difficult to implement these measures. For example, there is the lack of social protection systems, poor public and private infrastructure, the widespread informal job market and a large share of the population with low education levels. The chapter analyses the effects of the pandemic on the economy in the presence of a large informal sector. For the measures of physical distancing to take effect, it is necessary to implement public policies that provide subsistence conditions for individuals who are below the poverty line, i.e. those mostly working in the informal sector. Yet, the sheer size of the informal sector and the lack of social protection measures pose challenges to the fight against Covid-19 in developing countries. Consequently, the chapter provides, firstly, a reflection on the impacts of Covid-19 on the informal economy in developing and less developed countries, mainly analysing the public policies implemented for aiding the informal economy. Secondly, the chapter presents the expectations for the post-Covid-19 future. Finally, it presents policy recommendations to stimulate the recovery of the most affected countries.

7.1 Introduction

In developing countries, the informal economy is an outlet for workers who would otherwise be condemned to unemployment and extreme poverty. Indeed, most people enter the informal market out of necessity and not out of their own will.

M. Romanello (✉)
Department of Economics and International Relations, Federal University of Santa Catarina (UFSC) and International Institute of Social Studies (ISS), Erasmus University Rotterdam, Rotterdam, The Netherlands
e-mail: michele.romanello@ufsc.br

© The Author(s), under exclusive license to Springer Nature Switzerland AG 2022
E. Papyrakis (ed.), *COVID-19 and International Development*,
https://doi.org/10.1007/978-3-030-82339-9_7

The informal economy is an economic sector composed by a structural and a conjunctural component. In the case of structural component, workers and entrepreneurs join informality because they are individuals endowed with certain characteristics (poorly educated, women, young people, with backgrounds in the informal market, etc.). In most of developing countries, this structural component leads to a segmentation in the economy: individuals working in the informal sector are completely different from individuals in the formal sector, mainly in terms of their level of education and productivity, thus the transition of individuals from one sector to the other is very unlikely.

In the case of the conjunctural component, workers and entrepreneurs join informality depending on a country's situation: these individuals have skills and the educational level to belong to the formal sector, but in some cases, due to their economic situation (crisis, high unemployment, economic restructuring of some sectors, etc.), they must relocate to the informal economy to generate (additional) income and ensure their livelihood.

Although the current pandemic caused by Covid-19 is a conjunctural event, it affects both components of informality in developing countries.

Concerning the structural component, informal workers and entrepreneurs face a dilemma. If they follow the indications of the health authorities, with the possibility of lockdown, they lose a large part of their salary or profit. If they continue to work normally, they risk being infected by the virus, thus affecting their own health and their family members. In urban zones, even if they stay at home, these workers and their families have a high probability to be infected by the virus because of overcrowded and unhealthy living conditions that make social isolation almost impossible. Moreover, hygiene practices, like handwashing, are limited by the lack of piped water and soap, most of times, women are forced to collect water from outside of the home, thereby exposing themselves and their community (Mukhtarov et al., 2021).

Concerning the conjunctural component, the pandemic also further increases the number of individuals in informality: In a health and economic crisis, formal enterprises dismiss many workers or shut down completely and consequently, workers or entrepreneurs must transit to informality for the sake of survival (see also Demena et al., 2021 and Murshed, 2021).

7.2 The Impacts of Covid-19 on the Informal Economy in Developing and Less Developed Countries

Since the beginning of the Covid-19 crisis, it has been known that developing countries would be affected by two major shocks: one of sanitary origin and the other one of economic origin. The measures of social isolation and the closure and interruption of several economic activities, necessary to contain the spread of the epidemic in these territories, placed a huge portion of the economically active population at risk of unemployment, informality or caused significant losses of income.

Informal workers and the traditionally most vulnerable populations in these countries suddenly face a scenario of even greater economic and financial difficulties: acting outside the formal rules of the labor market, these people have no protection during periods of unemployment, loss of income or profit, and injury or illness (a possibility that increases a lot during the pandemic).

These facts are not a novelty of the current pandemic: in the previous health crises, those most affected belonged to the most vulnerable part of the population in developing countries, as described beneath.

In April 2009, a new strain of influenza H1N1 (swine flu) caused outbreaks in Mexico and the United States. The World Health Organization (WHO) declared it a pandemic in June 2009, since this new strain of the virus quickly spread across the world.

Smith and Keogh-Brown (2013) estimated the macroeconomic impact of H1N1 in Thailand, South Africa and Uganda. Their computable general equilibrium analysis of the pandemic (H1N1) indicates that the economic impact of that health crisis was small. However, some evidence suggests that the necessary absence from work caused by illness and death might have had a more negative impact on developing countries, especially for individuals with less social protection and in the informal sector compared with developed countries.

Analysing the socioeconomic impacts of another recent health crisis, Ebola (2013–2016) in West Africa, Fu et al. (2015) showed with nationally representative household surveys that hours worked dropped for those in informal employment and that non-farm household businesses were severely hit during the worst part of the crisis.

Between 2015-2017, the world observed a large health emergency in Brazil stemming from Zika virus. The emergency had a close link with the informal economy. On the one hand, the virus spread more easily in the suburbs of large cities due to the scarcity of social and health infrastructure. These areas are normally inhabited by people who work informally. Thus, their abysmal working and living conditions favoured the proliferation of the virus. On the other hand, the virus amplified the number of people working in the informal sector, mainly from two groups: women and black people. Although it was not possible to quantify the exact number of women and girls (compared to boys and men) who left school as well as the formal or informal labor market to care for a child or relative affected by Zika, a study of UNDP (2017) suggests that the burden fell disproportionately on women, which are largely present in the informal sector. According to data from the Brazilian Health Ministry (2017), in the context of the Zika virus epidemic, black families represent approximately 77% of confirmed cases. The situation produced a cycle of impoverishment by reducing the sources of income available, since the necessary care for children with congenital Zika syndrome implies an exit from the so-called productive work.

Compared to the earlier health crises and pandemics, the Covid-19 pandemic has some characteristics that make it unique.

The current pandemic shows echoes from the past. Elements common to earlier pandemics, such as panic and the appearance of fake news, can also be found now. However, we face the novelty of a global lockdown, being implemented by hundreds of countries at the same time. In developing countries, the social isolation resulting from the lockdowns have led to an increase in inequality: individuals with skilled jobs are normally allowed to work remotely at home, while unskilled workers must leave their home to work and are thus more easily prone to infection.

Similarly, the impact on schools is also a novelty when compared with past pandemics. In all countries of the world, Covid-19 has greatly reduced the annual hours of in-classroom teaching and consequently the quality of education of children and young people (Gómez & Andrés, 2021). And again, the consequences of school closures disproportionately affect developing countries. In turn, developed countries are better equipped and have more technological resources for online lessons. Furthermore, in most developed countries, the resources for online education are more or less equally distributed across primary, secondary and university education as well as across population strata and location. Yet, in developing countries these resources are normally exclusive to big cities and the rich or the elite.

A further distinguishing factor of the current pandemic is the massive use of digital means of communication. Covid-19 is the first health emergency in which we have an enormous amount of information spread across all countries of the world. This information is mixed: on the one hand, there are the recommendations of the World Health Organization and the rules of each country to combat the spread of the virus; on the other hand, there are countless fake news widely present in developing countries, which make it difficult for governments and international organizations to combat Covid-19, especially when the target audience of fake news are less educated population groups.

To begin the analysis of the effects of the pandemic on workers and entrepreneurs in the informal sector, it is of fundamental importance to look at the size of the informal sector in developing and less developed countries.

In what follows, I present data about the size of the informal sector in the economy across different developing countries and their development over time. Table 7.1 presents the informal employment rate in terms of percentages for the period 2010 to 2019. These data are harmonized and use the same set of criteria to measure informal employment across countries to assure comparability.

According to the presented share in employment, informal employment varies considerably between the countries considered: it goes from about 30% in Bosnia and Herzegovina and Uruguay to a value of over 90% in Ivory Coast, Mali and Togo. Strikingly, over time, we do not observe large variations or declines in the shares of informal employment within countries highlighting that informality is a strong structural component across developing countries. Only Albania, Armenia, Niger, Thailand, Uruguay and Vietnam have managed to reduce employment in the informal sector by 10 percentage points or more over the 10 years (or less) considered in Table 7.1.

The informal employment shares presented in Table 7.1 are useful for categorizing developing countries into two groups: countries with high informality (with a

7 Covid-19 and the Informal Sector

Table 7.1 Informal employment rate across developing countries

	2010	2011	2012	2013	2014	2015	2016	2017	2018	2019
Albania			66.6	61	65.6	63.4	61.6	58.7	57.9	56.7
Argentina	48.5	47.7	47.8	47.4	46.8			47.9	48.1	49.4
Armenia		62.4	61.5	60.4	59.1	52.1	49.9	47.9	50.3	
Bangladesh	87.8			89.4				94.7		
Bolivia		83.3	82.1	80.4	85.2	84.1	76.3	79.1	80.7	84.9
Bosnia and Herzegovina	32.4	30.9	33.7	31.5	29.3	30.1	30.5	24.8	22.1	23.1
Brazil		46	45.1	44.1	44.8	45				
Brunei Darussalam					33			37.7	37.5	30.9
Colombia	67.9	67	66.6	64.9	63.2	63.1	62.4	61.9	62.4	62.1
Costa Rica	37.9	34.7	36.8	38.8	38.8	39.7	37.9	38.2	37.8	38.8
Ivory Coast				94.8	97.7			92.8	86.7	
Dominican Republic	53.9	56.2	56.3	55.9	54.2	56.3	56.3	57.2	56.8	54.3
Ecuador	75.5	72.6	71.1	69.7	67.6	68.2	70.8	72.4	72.7	73.6
Egypt		58.7	55.9	57.6	59.8		61.9	63.8	63.4	
El Salvador	72.5	72.6	72.6	71.4	69.3	69.1	69.5	70.2	68.5	69.1
Gambia			76.5						77.6	
Ghana				92.1		88.8				
Guatemala	80.9	81.2	82.7	80.5	78.1	79.2	79.2			
Honduras	80.4	81.9	83.7	93	91.7	88.8	90.2	82.6		
India	87.9		86.5						88.6	
Indonesia							84.3	84.2	82.4	80.4
Mali				89.9	95.3	96.1	94.9	94.7	93.4	
Mauritius			51.3	51.5	54.7	53.3	52.9	52.2	53.5	53.5
Mongolia	52.1	56.3	58.6	56.3	49.8	50	50.7	51.3		48.2
Namibia				54.2	59.4			66.9	55.8	
Niger		95.4						78.2		
Pakistan	81.1	82.4		81.6	83.3	82.5		81.8		
Panama		46.5	47	47.9	48	47.8	48.5	49.4	51.4	52.8
Paraguay	76.7	73.9	74.2	71.5	70.1	70.7	72.2	71.3	70.3	68.9
Peru	76.9			69.8	68.6	69.2	68	68.1	68.5	68.4
Rwanda								80.4	80.4	80.1
South Africa	56.7	54.5		54	53.2	55.6	54.9	54.2	54.9	54.1
Sri Lanka		70.8		70.8	69.9	69.6	69.7	68.2	68.1	
Thailand					75.8	74.6	75.1	67.2	64.4	
Togo					95.1			90.1		
Uganda			91.7					89.4		
Uruguay	40.9	38.3	36.2	35.4	23.6	23.9	24.5	24.1	24	24
Vietnam				79.5	76.9	75.5	74.1	73.3	71.4	67.3
Zimbabwe			85.6			87.4				79.8

Source: ILOSTAT – International Labour Organization

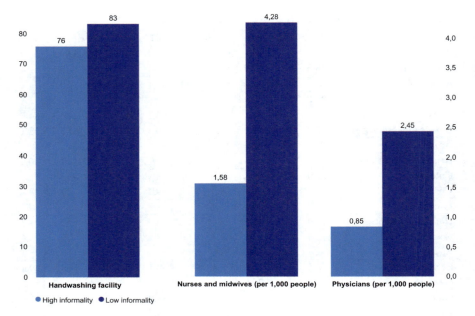

Fig. 7.1 Access to handwashing and medical staff by formality group. (Source: Own elaboration using data from: World Bank, World Development Indicators; WHO/UNICEF Joint Monitoring Program (JMP) for Water Supply, Sanitation and Hygiene; WHO)

share greater than or equal to 60%) and low informality (below 60%). Applying this classification, I can assess the level of resources to prevent and combat Covid-19 available to developing and less developed countries depending on the level of informality.

Figure 7.1 shows the share of the population that has access to handwashing facilities, the average number of nurses and midwives available per 1000 people and the average number of physicians available per 1000 people. The light blue bars in Fig. 7.1 code for the group of countries with a high level of informality and the dark blue bars for those with low informality.

High levels of informality are associated with limited access to medical resources, as well as hygiene facilities. People from countries with high informality have a lower access to handwashing facilities; the difference is 7 percentage points compared to countries of low informality. Moreover, when medical staff is considered the difference between the two groups becomes even bigger: the disparity is 2.7 nurses and midwives and 1.6 physicians per 1000 people respectively.

Other important data to be considered are the number of workers affected by the pandemic according to whether they work in the formal sector or not. As outlined in the introduction, in a pandemic, the fundamental difference between the two types of workers is that informal workers need to work to receive an income (and sometimes to survive) even though they can be infected, while often formal workers either receive their salary even if they do not work or they have the possibility of home working or they are protected by welfare schemes.

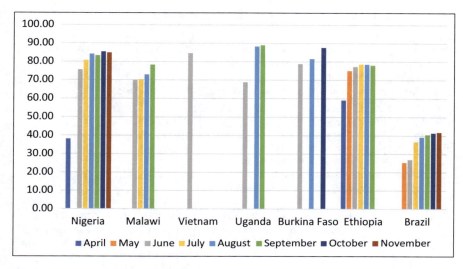

Fig. 7.2 Percentage of individuals who did some work, business or farming in the period April to November 2020. (Source: Own elaboration using data from: World Bank Covid-19 Phone Surveys of Households and Brazilian National Household Sample Survey-Covid-19)

Figure 7.2 presents the percentages of individuals that did some work for pay, some kind of business, farming or other activity to generate income, even if only for 1 h a week, from April 2020 to November 2020. The data is only available for a small set of selected countries, namely Nigeria, Malawi, Vietnam, Uganda, Burkina Faso, Ethiopia and Brazil.

All countries for which data are available show a comparably low share of individuals working normally in the first months of the pandemic; however, this share is increasing in the following months. This shows that workers or entrepreneurs were initially caught off guard or were profoundly afraid of the pandemic. Yet, as the months passed the need to earn income and to come up for their livelihood has gained weight. The country where this observation is most evident is Nigeria: the share of those doing some work went from 38% in April 2020 to an average of 80% in the subsequent months.

An additional aspect to be analysed is the existence and type of policy implemented to help informal workers. Workers of the informal sector are among the main populations targeted by emergency support. Social assistance schemes can be implemented as a policy tool to help these families make ends meet and to keep these people at home in the worst moments of the pandemic.

However, only some countries implemented policies targeted specifically towards informal workers, the majority created social assistance for vulnerable households in general. Table 7.2 shows the countries that implemented policies for informal workers and entrepreneurs, the type of policy and the value the received per month as emergency support.

Table 7.2 Policies for informal workers and entrepreneurs affected by pandemic

Country	Type of policy	Value per month
Albania	Informal self-employed families receive a special benefit equivalent to a state-set monthly salary	22,000 Albanian Leks (USD 214)
Argentina	A lump sum payment to one member of the family	$10,000 (USD 155)
Brazil	Emergency cash transfer for 6 months	USD 115
Burkina Faso	Suspending government fees charged on informal sector operators for rent, security and parking in urban markets	Suspension of government fees
Cabo Verde	One-time cash transfer for self-employed individuals in the informal sector	10,000 Escudos (USD 100)
Chile	A cash transfer program for 3 months intended for households that depend on informal work for their livelihood	CLP$ 280,000 (USD 340) for first payment. First pay-out corresponds to 100 percent of the full transfer amount, the second pay-out 85 percent and the third 65 percent
Colombia	One-time cash transfer for informal workers	COP 160,000 (USD 108)
Costa Rica	A compensation for the fall in the income of informal workers	125 k Colones (USD 223)
Ecuador	Cash transfer for informal workers who earn less than USD 400 per month	USD 60
Egypt	One-time cash transfer	EL 500 (USD 32)
El Salvador	One-time cash transfer to households who work in the informal economy	$300 (USD 34)
Guatemala	Emergency cash transfer – Eligibility based on electricity consumption	GTQ 1000 (USD 130)
Indonesia	Vouchers for training and re-skilling	Rupee 3.5 million (USD 223)
Lesotho	Expanded social protection programs to include informal sector workers	42.500 Tenge (USD 102)
Malaysia	One-time grant program for SMEs	MYR 3000 (USD 690)
Morocco	A mobile payment to workers operating in the informal sector	−800 dirhams(USD 89) per month for households of two people or less −1000 dirhams(USD 112) per month for households of three to four people −1200 dirhams (USD134) per month for households of more than four people
Namibia	One-time cash transfer to people who have lost their jobs, either in the informal or formal sector	N$ 750 (USD 49)
Nepal	Food assistance package distributed to informal sector laborers through ward committees	Food assistance package
North Macedonia	Expanded unemployment insurance system to 20,000 households	7000 denars (USD 125)

(continued)

Table 7.2 (continued)

Country	Type of policy	Value per month
Philippines	Payments for 2 months to households working in the informal economy	P 5000 – P 8000 (USD 102 - 164) a month
Rwanda	In-kind distribution of food and other essential items to casual laborers whose livelihoods depend on daily wage and self-employed mainly in the informal sector	In-kind distribution of food and other essential items
Saint Lucia	Monthly payment for 3 months, preconditioned on being enrolled in National Insurance Corporation for informal workers	$500 (USD 185)
Sierra Leone	Emergency cash transfers to vulnerable households with informal sector workers in major cities for a period of 9 months	From USD 15 to USD 30
Sudan	Support for informal workers in the form of food basket with five commodities	3000 SDG (USD 55) each food basket
Thailand	Cash transfer for 3 months for informal workers	5000 Baht (USD 153)
Togo	State grant for informal workers of at least 30 percent of the minimum wage	CFAF 10,500 - 20,000 ((USD 18 - 34)
Tunisia	One-time cash transfer to households working in the informal sector	TND200 (USD 68)
Vietnam	Expanded social protection programs to include informal sector workers	VND one million (USD 44) for each poor household, VND 500,000 (USD 22) per month for each near-poor household

Source: Gentilini et al. (2020), Nygaard and Dreyer (2020)

According to Table 7.2, cash transfer programs for informal workers and entrepreneurs tend to be of short duration but relatively generous in value. Another aspect that can be deducted from this Table is that the interventions were generous in the first months of the pandemic and then faded over time, leaving informal workers uncovered in recent months. This observation also supports the findings presented in Fig. 7.2 that people quickly had to resort to their (informal) productive activities to cover their expenses.

Gentilini et al. (2020) estimated that 136.7 million informal workers have received cash transfers from some government program in 2020. This is likely a conservative estimate, since there are other general programs to aid vulnerable households which also affected the informal workers positively. Yet, overall the short-lived nature of the support and the limited means the cash-constrained governments of most developing countries have at their disposal forced the people to take up their activities quickly again even risking an infection with Covid-19.

7.3 Expectations for the Post-Covid-19 Future

The Covid-19 pandemic has had unprecedented impacts on the labor market and, therefore, it is difficult to make predictions about post-Covid-19 developments. But pre-pandemic trends and previous crisis provide some indications.

In the coming months, there will probably be an increase in competition in the labor markets in developing countries. This will mainly be triggered due to the decrease in the number of vacancies and likely economic recessions and even due to the agreements that some workers achieved with their employers to preserve their jobs. In such situations, it is common for companies to value more experienced workers, which can be a hurdle for younger workers.

Another aspect that has further been spurred by the pandemic may be the reduced need for less skilled workers. Many jobs cut during social isolation are at risk of not being re-opened. Before the pandemic, the forecast was that by 2030 about 14% of the global workers would have to change jobs because of automation (Manyika et al., 2017). With the pandemic and the advancement of digitalization, this transition period is likely shortened.

Some sectors were positively affected by the pandemic and it is likely that they will maintain this advantage also in the post-Covid-19 period. Sectors that can be included in this group are those that are technology heavy or depend on the internet. For example, app companies, which deliver food, benefitted from the pandemic. Similarly, after experimenting with online shopping, the population in many developing countries broke the resistance that existed towards these digital services. Thus, programming might offer new alternatives for workers whose employment contracts were terminated. Yet, it requires quite some skills and training and is thus not necessarily an option for individuals in the informal sector. The implications of these developments will be most noticeable on the labor market when the economy returns to growth and many unskilled workers are unable to relocate to their original areas.

However, the expansion of these sectors may lead to a revaluation of broadly utilised employment practices that tend to diminish workers'conditions and health and safety protection, with the intention of obtaining a competitive advantage. The intensive use of the "gig-economy" (people working in short-terms jobs, often being self-employed) by app companies may grow with a larger need for flexibility among employers and, consequently, this will aggravate the size of the informal sector. Most probably, their aversion to taking on staff may be further exacerbated by the possibility of potential health and economic disruptions in the near future (Webb et al., 2020).

Another aspect that is interesting to consider is the expansion of Industry 4.0 in the next years and its effects on informality in developing countries.

In a study presented in 2017, the consultancy company McKinsey shows that about 800 million professionals could lose their jobs by 2030. The report analysed 800 professions in 46 countries and found that up to a third of the current jobs could be automated in 12 years. In countries of central capitalism, such as the United

States and Germany, between 23% and 24% of the current jobs will suffer directly from automation according to this analysis. It is further argued that peripheral countries, which have less money to invest in automation and robotics, would not be affected very much until 2030. For example, in India, the impact would be only on 9% of the jobs, while in Brazil, this share could reach 15% (Manyika et al., 2017).

This points to a large structural transition, namely that developing countries will be partially excluded from world production and a process of production transfer will take place in the reverse direction compared to the one that occurred in the last decades. Put differently, the migration of manufacturing firms to countries with cheaper labor will be replaced by the relocation of production to countries that intensively use robots and automations. For example, this development can already be observed with telemarketing: in the last decade a job transfers to India could be observed because wages are cheaper than in the USA and Europe. But currently we witness that this work can be done by a robot.

Also, inside developing countries, a restructuring of the production sector will take place due to the automation of traditional manufacturing and industrial practices. For example, in Brazil there is a replacement of cane cutters by mowers since the latter work 24 h a day, replace up to 15 workers, and are more efficient. Thus, the need for cutters in the production process is no longer necessary (Evangelista, 2018). Thus, based on pre-pandemic automation processes it can be expected that informal workers and entrepreneurs will be heavily affected by industry 4.0 in the years to come.

There is one sector that is likely to have a large demand for workers after the pandemic, the health sector. Yet again, more than basic education is required to enter the sector and it is not an alternative for older informal worker.

Thus, the pandemic has likely reinforced and sped up already existing trends that make it harder and harder for informal sector workers.

7.4 Policy Recommendations for Interventions in the Informal Sector

As the word itself implies, the informal sector is a hidden sector that is difficult to identify by policymakers. Moreover, this sector is heterogeneous (Floridi et al., 2016): in most countries it is composed by a structural and a conjunctural component, as outlined in the introduction. Consequently, public policies, both under normal conditions and in the condition of a health and social emergency must be adapted to the different characteristics, conditions and needs of the workers involved.

However, there are some recommendations that are universal for all developing countries. The first policy recommendation is the obvious need for greater investments in the health systems of developing countries. The Covid-19 crisis showed the failure of countries that neglected their public health systems during decades of austerity policies. The importance of allocating greater expenditures to public health systems is evident. The consequence is that the financing of the expected health expenditures will further raise the debt of the current health care and social security systems.

However, the fact that the investment in health care, like in basic education and in other crucial elements of social opportunity, is highly labor-intensive makes it affordable even in weaker economies where labor is cheaper. Importantly, it is likely to have a positive impact on the labor market (Sen, 1999).

In additional, equal access to vaccines will be essential to overcome the crisis, but, even more, to avoid an increase in economic and social inequality in the world and to avoid pushing even more people into informality. In this regard, by January 2021, more than 180 countries have joined the COVAX initiative. Its objective is to unite countries worldwide in the vaccination efforts and to support in particular developing countries so that they have greater power in the negotiations with the pharmaceutical companies. Was COVAX not set up, there is the real threat that most people in the world would not be immunized against SARS-CoV-2, a burden that would be disproportionately shouldered by the poor and the informal sector. COVAX is necessary to guarantee that ability to pay does not become a barrier to accessing vaccines (Okonjo-Iweala, 2021). However, the mere creation of an initiative to offer vaccine doses in an equitable manner does not guarantee the effectiveness of vaccination campaigns in developing countries. There are other issues to be considered: internal inequalities, the heterogeneity of these countries, the accessibility of remote areas, the availability of health professionals and medical inputs (such as syringes), and the governments who really want to vaccinate their population.

On this matter, Burns Loeb et al. (2021) show how connectedness, i.e. internet access, is emerging as a new and problematic determinant of health. This seems to be especially true for cash-constrained minority communities. It is more practical and faster to book a vaccination appointment online than by calling. Consequently, internet access is increasingly seen as a fundamental civil rights issue. However, expanding internet access is a long-term investment. To address the problem in the short-term, especially during this pandemic, policymakers must identify where internet access is potentially a barrier and protect against negative repercussions. This policy could involve reserving vaccines for minority communities who tend to constitute a large share of the informal sector.

Another policy recommendation is the implementation or extension of income security policies, in particular for informal sector workers. Effective policies to improve income security for workers in the informal economy are necessary across developing countries, especially when we consider families composed by female-headed households with children. There is a need to expand public social security systems to include workers in the informal sector. In addition to income support policies, developing countries should put in place policies for food support.

Moreover, fiscal and monetary measures can be developed to help workers and entrepreneurs in the informal economy. As suggested by the ILO (2020), financial support can be supplied to cover the debts of families and firms. They can take the form of grants, subsidies, and/or extensions of debt periods. Moreover, waivers or delayed payments for public services, such as electricity, water or rent, can also be authorized with the aim of decreasing operating costs. Furthermore, the internet must be extended to all social classes, along with training, in order to enable some units in the informal economy to experiment with digital tools for business continuity, creation of new opportunities and income generation. It is especially noteworthy the expansion of remote work from home for all workers regardless of income level.

In respect of informal businesses, previous evidence recommends that policymakers should focus on interventions that increase the benefits of formalization and, if possible, attempt to develop new policies that go beyond solely aiming for the formalization of informal firms. The policies could include the strengthening of the existing links between the formal and informal economy. In this viewpoint, including informal entrepreneurs in the policymaking process could be a practicable choice for developing specific policies and for elaborating innovative strategies addressed to informal firms (Floridi et al., 2020).

In respect of informal workers, an obstacle to formalization could be the unresolved tensions between informal workers'desire for job stability on the one hand and, on the other, some employers' determination for continued labor flexibility started during pandemic. The contrast between employer-worker perspectives leads to the employers' effort to transfer the risk/costs of such flexibility to government and workers. Therefore, it is essential to analyse how government policies, including labor market control and assistance for continued employment during a pandemic, can resolve this contrast in the longer-term (Webb et al., 2020).

Finally, long-term investments in the educational systems of developing countries are fundamental to stimulate the transition of workers and informal entrepreneurs to formality. In the case of structural informality, education provides individuals with the skills and knowledge to increase productivity. In the case of conjunctural informality, education offers some skills for relocation to the formal labor market in the case of job loss due to crisis, as currently experienced due to Covid-19.

References

Brazilian Health Ministry. (2017). *Saúde Brasil 2015/2016. Uma análise da situação de saúde e da epidemia pelo vírus Zika e por outras doenças transmitidas pelo Aedes aegypti*. Retrieved from: https://portalarquivos2.saude.gov.br/images/pdf/2017/maio/12/2017-0135-vers-eletronica-final.pdf (Last accessed: December 28, 2020).

Burns Loeb, T., Adkins-Jackson, A. J., & Brown, A. F. (2021). *How lack of internet access has limited vaccine availability for racial and ethnic minorities*. World Economic Forum. Retrieved from: https://www.weforum.org/agenda/2021/02/internet-vaccine-racial-ethnic-minorities-covid-coronavirus-pandemic?utm_source=facebook&utm_medium=social_scheduler&utm_term=COVID-19&utm_content=24/02/2021+14:30 (Last accessed: February 25, 2021).

Demena, B. A., Floridi, A., & Natascha, W. (2021). The short-term impact of COVID-19 on labour market outcomes: Comparative systematic evidence. In P. Elissaios (Ed.), *Covid-19 and International Development*. Springer.

Evangelista, A. P. (2018). *Seremos líderes ou escravos da Indústria 4.0? EPSJV/Fiocruz*. Retrieved: http://www.epsjv.fiocruz.br/noticias/reportagem/seremos-lideres-ou-escravos-da-industria-40 (Last accessed: January 23, 2021).

Floridi, A., Wagner, N., & Cameron, J. (2016). *A study of Egyptian and Palestine transformal firms – A neglected category operating in the borderland between formality and informality*. ISS Working Paper Series/General Series 619, 1–25. Retrieved from: https://ideas.repec.org/p/ems/euriss/80085.html (Last accessed: March 10, 2021).

Floridi, A., Demena, A. B., & Wagner, N. (2020). *Shedding light on the shadows of informality: A meta-analysis of formalization interventions targeted at informal firms*. Labour Economics 67, 101925. Retrieved from: https://www.sciencedirect.com/science/article/pii/S0927537120301299 (Last accessed: March 15, 2021).

Fu, N., Glennerster, R., Himelein, K., Rosas, N., & Suri, T. (2015). *Socioeconomic impacts of Ebola: Sierra Leone, Round 2*. World Bank Report, World Bank, Washington DC. Retrieved from http://www.worldbank.org/en/topic/poverty/publication/socio-economic-impacts-ebola-sierra-leone (Last accessed: December 22, 2020).

Gentilini, U., Almenfi, M., Dale, P., Lopez, A.V., & Zafar, U. (2020). *Social protection and jobs responses to COVID-19: A real-time review of country measures. Living paper version 12*. World Bank. Retrieved from: http://documents1.worldbank.org/curated/en/454671594649637530/pdf/Social-Protection-and-Jobs-Responses-to-COVID-19-A-Real-Time-Review-of-Country-Measures-July-10-2020.pdf (Last accessed: March 02, 2021).

Gómez, G. M., & G.J. Andrés Uzín P. (2021). Effects of COVID-19 on education and schools' reopening in Latin America. In E. Papyrakis (Ed.), *Covid-19 and International Development*. Springer.

ILO. (2020). *COVID-19 crisis and the informal economy. Immediate responses and policy challenges*. Briefing Note, International Labour Organization. Retrieved from : https://www.ilo.org/wcmsp5/groups/public/%2D%2D-ed_protect/%2D%2D-protrav/%2D%2D-travail/documents/briefingnote/wcms_743623.pdf (Last accessed: January 18, 2021).

Manyika, J., Lund, S., Chui, M., Bughin, J., Woetzel, J., Batra, P., Ko, R., & Sanghvi, S. (2017). *Jobs lost, jobs gained: What the future of work will mean for jobs, skills, and wages*. McKinsey Global Institute. Retrieved from: https://www.mckinsey.com/featured-insights/future-of-work/jobs-lost-jobs-gained-what-the-future-of-work-will-mean-for-jobs-skills-and-wages# (Last accessed: February 03, 2021).

Mukhtarov, M., Papyrakis, E., & Rieger, M. (2021). Covid-19 and water. In E. Papyrakis (Ed.), *Covid-19 and International Development*. Springer.

Murshed, S. M. (2021). Consequences of the Covid-19 pandemic for economic inequality. In E. Papyrakis (Ed.), *Covid-19 and International Development*. Springer.

Nygaard, K., & Dreyer, M. (2020) *Countries provide support to workers in the informal economy*. Blog. Yale School of Management. Retrieved from: https://som.yale.edu/blog/countries-provide-support-to-workers-in-the-informal-economy (Last accessed: March 01, 2021).

Okonjo-Iweala, N. (2021). *Globalizing the COVID vaccine*. GAVI. Retrieved from: https://www.gavi.org/vaccineswork/globalizing-covid-vaccine (Last accessed: January 21, 2021).

Sen, A. (1999). *Economics and health*. The Lancet 354, Special Issue, SIV20. Retrieved from: https://doi.org/10.1016/S0140-6736(99)90363-X (Last accessed: March 03, 2021).

Smith, R. D., & Keogh-Brown, M. R. (2013). Macroeconomic impact of a mild influenza pandemic and associated policies in Thailand, South Africa and Uganda: A computable general equilibrium analysis. *Influenza and Other Respiratory Viruses, 7*(6), 1400–1408. Retrieved from: https://doi.org/10.1111/irv.12137 (Last accessed: January 11, 2021)

UNDP. (2017). *Uma avaliação do impacto socioeconômico do vírus Zika na América Latina e Caribe*: Brasil, Colômbia e Suriname como estudos de caso. United Nations Development Programme, New York. Retrieved from: https://www.br.undp.org/content/dam/rblac/docs/Research%20and%20Publications/HIV/UNDP-RBLAC-Zika-07-20-2017-Portuguese-WEB.pdf (Last accessed: January 3, 2021).

Webb, A., McQuaid, R., & Rand, S. (2020). Employment in the informal economy: implications of the COVID-19 pandemic. *International Journal of Sociology and Social Policy, 40*(9/10), 1005–1019. Retrieved from: https://doi.org/10.1108/IJSSP-08-2020-0371 (Last accessed: March 14, 2021)

Chapter 8
Indirect Health Effects Due to COVID-19: An Exploration of Potential Economic Costs for Developing Countries

Natascha Wagner

Abstract Modelling estimates infer almost 1.2 million indirect COVID-19 deaths in developing countries from additional (i) child and maternal deaths, (ii) HIV, (iii) tuberculosis, and (iv) malaria, corresponding to almost 50% of the 2.5 million direct COVID-19 deaths worldwide reported by late February 2021. Furthermore, indirect victims of the pandemic are expected from chronic diseases, cardiovascular problems, cancer, and neurosurgical conditions as well as long-term health consequences from ophthalmic diseases, dental care needs, hookworms, bacterial infections and others. Malnutrition and obesity are also likely to be exacerbated and additional mental health needs to arise in the already overburdened, poorly equipped and underfinanced developing country health systems. Next to identifying the burden of disease that is largely neglected because of COVID-19, this chapter provides exploratory analysis of the economic costs of delayed and foregone health care. This shows that: a) in the worst case, these costs can hamper the already low economic growth in the Sub-Sahara Africa region and b) that developed countries do good in supporting developing countries in their broader health system responses, even if the main objective of the former is to avoid further Coronavirus mutations and their spread across the globe.

8.1 Introduction

Due to COVID-19 the world has faced more than 83 million recorded infected individuals and more than 1.8 million direct deaths by the end of 2020. Despite the first vaccines appearing on the horizon and being officially authorized, in early 2021 new virus mutations further exacerbated the spread of the virus. In February 2021 the total number of COVID-19 cases exceeded 110 million and directly related

N. Wagner (✉)
Department of Development Economics, International Institute of Social Studies, Erasmus University Rotterdam, The Hague, The Netherlands
e-mail: wagner@iss.nl

© The Author(s), under exclusive license to Springer Nature Switzerland AG 2022
E. Papyrakis (ed.), *COVID-19 and International Development*,
https://doi.org/10.1007/978-3-030-82339-9_8

deaths surpassed 2.5 million worldwide. Within only 2 months – between December 31st, 2020 and February 28th, 2021 – the number of deaths increased by more than 700,000 globally or almost 40%. The direct health crisis is undeniable and its imminent threat still needs to be banned. Beyond closures of shops, enterprises, the catering trade, and schools the directly lost lives have huge economic implications. It is estimated that as of January 6, 2021, the years of life lost due to COVID-19 are 2–9 times the average seasonal influenza in severely affected countries and worldwide 20.5 million years of life (Pifarré i Arolas et al., 2021). If only attributing a value of 100 US$ to every year of life, the direct global loss due to COVID-19 is tremendous.

Yet, this is by no means the end of the story. A narrow focus on the direct costs associated with COVID-19 disguises the so far largely unaccounted for indirect health costs. As governments prolong lockdown measures, entrepreneurs keep losing their businesses, and school children are forced to stay out of school, we increasingly also observe considerable indirect health consequences from delayed health care seeking, deferred surgeries, unattended chronic diseases, undetected non-communicable diseases such as cancer, home deliveries without skilled birth attendants, as well as untreated mental diseases and lack of emotional wellbeing due to isolation and fear of losing one's livelihood.

In a first step, this chapter sets out to identify the indirect health effects resulting from the measures that have been put in place to curtail COVID-19 in developing countries. This includes delayed and foregone health care seeking and postponed treatment. In a second step the chapter presents an exploratory analysis of the economic costs associated with the decline in non-COVID related hospital admissions due to the pandemic. This analysis rests on hospital discharge data for Central and Eastern Europe and also provides exploratory calculations for Sub-Sahara Africa to the extent that data are available. Finally, the chapter discusses the health system implications for developing countries and concludes with two short-term responses for mitigating the indirect health impacts of COVID-19 for the populations in developing countries.

8.2 Indirect Health Effects Due to COVID-19 in Developing Countries

Early modelling exercises about the indirect effects of the Coronavirus pandemic on maternal and child health in low- and middle-income countries suggest 24,400 additional maternal and 506,900 additional child death due to wasting, pneumonia, diarrhea, lack or wrong application of antibiotics and unattended delivers over a 12-month period based on the assumption of relatively mild cutbacks in health care provision (Roberton et al., 2020). Home birth without skilled birth attendants are one of the reasons for additional neonatal deaths but also for long-term complications and handicaps. The implications for babies and mothers are so large since in the first 9 month into the pandemic an estimated 30 million children were born only

8 Indirect Health Effects Due to COVID-19: An Exploration of Potential Economic...

in the high-fertility countries which concomitantly tend to have high neonatal mortality rates, namely 20.1 million birth in India, 6.4 million in Nigeria and five million in Pakistan (UNICEF, 2020). A different yet related risk of the pandemic is the increase in teenage pregnancies (in combination with an increased likelihood of child marriage) across developing countries (UNICEF, 2021). To date, reliable data are not yet available to assess how many additional teenage pregnancies resulted from school closures, economics pressures on family and lockdowns.

In addition to the fertility related challenges faced by many developing countries there is the pre-existing, endemic disease burden. Already prior to COVID-19 Sub-Sahara Africa faced the triple burden of disease from HIV, tuberculosis, and non-communicable diseases. Voluntary HIV counselling and testing form an important ingredient into combatting HIV and aids. Yet, fear of Coronavirus infections and COVID-related stigma prevent people from accessing HIV test facilities (Lagat et al., 2020; Ponticiello et al., 2020). Moreover, due to lockdowns and disruptions in supply chains there is the risk of stock outs in antiretroviral medication (WHO, 2020a). It has been modelled that disruptions in the delivery and provision of antiretroviral medication over a period of six-months could result in 230,000 to 500,000 extra deaths (Jewell et al., 2020; WHO, 2020b).

The second disease that is threatening developing countries and especially Africa is tuberculosis. According to the WHO tuberculosis is the ninth leading cause of death worldwide with over 25% of tuberculosis' deaths occurring on the African continent (WHO, 2020c). A disruption in tuberculosis prevention, surveillance, and treatment programmes has been documented during the pandemic (Jain et al., 2020) and the imminent risk is that undetected and untreated cases further spread the disease (Saunders & Evans, 2020). The challenge faced by developing countries is that tuberculosis patients are more likely to contract other diseases such as HIV; a larger susceptibility to COVID-19 is also likely which might worsen the health condition of tuberculosis patients further (WHO, 2020d). According to WHO's tuberculosis report for 2020, a sharp decline in tuberculosis notifications can be observed in countries with a high disease burden. According to first estimates a 50% decrease in tuberculosis case detection over a period as limited as 3 months could result in 400,000 additional tuberculosis deaths (WHO, 2020e).

As in the case of HIV and tuberculosis prevention and treatment, COVID-19 also threatens the delivery of malaria services. Again, the African region is most disproportionately affected; the continent was home to 94% of the malaria cases and deaths in 2019 (WHO, 2020e). According to Hussein et al. (2020) disruptions in the distribution of insecticide-treated nets (ITNs), malaria chemoprevention and indoor residual spraying hold the potential to lead to serious malaria outbreaks as an indirect consequence of the pandemic. Based on evidence from the latest Ebola outbreak (2014–2016) in West Africa, a considerable increase in malaria-related illness and death due to COVID-19 can be expected (Walker et al., 2015). Put differently, WHO (2020f) modelled nine different scenarios of the fight against malaria during the COVID-19 pandemic. In the relatively mild scenario where insecticide treated net (INT) campaigns are disrupted and the distribution of the nets is reduced by

75%, an expected additional 30,000 lives would be lost to malaria compared to the baseline situation from 2018.

Taken together, the early and most conservative modelling exercises suggest almost 1.2 million additional, indirect COVID-19 deaths in developing countries only considering (i) child and maternal mortality, (ii) HIV, (iii) tuberculosis, and (iv) malaria. This corresponds to almost 50% of the 2.5 million direct COVID-19 deaths worldwide reported by late February 2021. While these numbers are shocking, it has to be acknowledged that the modeling exercises along with most of the health-related literature about the indirect effects of COVID-19 represent immediate responses to the pandemic. Until now, these estimates cannot be substantiated. The actual database to assess indirect effects is still fairly limited or rather inexistent, in particular for developing countries.

Although precise figures are lacking there is no doubt about more indirect health consequences stemming from COVID-19. First, the treatment of other chronic disease such as diabetes is considerably impaired in developing countries (Nouhjah & Jahanfar, 2020). Second, the diagnosis of cardiovascular diseases has seen a tremendous reduction across the world in response to COVID-19 raising the concern of a long-term worsening of cardiovascular health outcomes (Einstein et al., 2021). The same applies for cancer detection and treatment (Del Pilar Estevez-Diz et al., 2020; González-Montero et al., 2020). Similarly, neurosurgical patients with a stroke and/or brain tumors face severe treatment challenges since care systems for them have broken down (Dhandapani & Dhandapani, 2020). Developing country patients with these conditions are indirect victims of COVID-19. Third, the suspension of elective surgeries has also been observed in developing countries (Sarin, 2020). Yet we know hardly anything about cost implications for the hospitals and the concerned individuals (costs of foregone earnings and long-term health consequences). Fourth, the so-far established disease list leaves out long-term health consequences from patients with ophthalmic diseases, dental care needs, hookworms, bacterial infections and other diseases that can be cured if attended to. Not only is it very likely that people do not seek treatment, in Malaysia a case has been reported of a family that absconded from hospital due to fear of COVID-19 (Koh Boon Yau et al., 2020). Fifths, the discussion so far excludes the double burden of the nutrition transition that has been witnessed across developing countries where malnutrition and obesity co-exist (Mai et al., 2020; Gupta et al., 2012; Popkin et al., 2012) and likely have increased as a consequence of the pandemic. Moreover, observations of early coping mechanisms suggest that the behavioral responses of people in developing countries further exacerbate the health crisis as they resort to traditional therapies and over-the-counter medicines for self-treatment (Arthur-Holmes et al., 2020). Even under the best circumstances where we do not observe any deaths from the mentioned, largely unattended to medical conditions, productivity losses due to late treatment, wrong self-medication and/or long-term health impairments are likely and will lead to sizable deficits in national output.

In addition, mental health needs arise as a consequence of the pandemic itself and measures to contain the spread of the virus. Mental health implications of COVID-19 for developing country populations have not yet been identified in much detail, in particular representativeness is lacking (Haider et al., 2020). Yet, a study about the immediate psychological responses to COVID-19 in the general Chinese population suggests that more than half of the 1210 respondents from 194 cities rated the psychological impact as moderate-to-severe, and about one-third reported moderate-to-severe anxiety (Wang et al., 2020). For the case of Nepal, there is evidence that mental, spiritual and social wellbeing went down in response to curfews, self-isolation, social distancing and quarantine (Poudel & Subedi, 2020). Evidence from India points out that already during the initial stages of the pandemic, almost one-third of the more than 1000 respondents to an online survey reported a significant psychological impact (Varshney et al., 2020). Furthermore, a study among children in confinement in Hubei province, China suggests increased depressive symptoms and anxiety (Xie et al., 2020). In addition, there is the report of a hospital suicide in Bangladesh due to treatment neglect during the pandemic (Mamun et al., 2020). Most of these studies focus on immediate impacts and have small samples that suffer from self-selection. Long-term impacts are unfolding as this article is written and their implications and costs still need to be identified.

At the same time there is a large awareness of the psychological distress that health care workers are going through. During the early stages of COVID-19, Zhang et al. (2020) presented evidence of increased psychosocial problems among medical health workers in China suggesting that these might need long-term attention. Evidence from Nepal further indicates that 38% of the health care workers on COVID-19 duty are suffering anxiety and/or depression (Gupta et al., 2020). Yang et al. (2020) call for psychological interventions for frontline workers, particularly in low- and middle-income countries. Among other support mechanisms e-mental health solutions for health care workers have been tested and are advocated for (Drissi et al., 2020). Moreover, it is know from psychological research conducted in the aftermath of Ebola that fear-related behaviours and stigmatisation are common and that post-traumatic stress symptoms are most likely among those who had exposure to the disease (O'Leary et al., 2018). Thus, mental health struggles among frontline health care workers have the potential to put stress on the health care system in the years to come.

Last but not least, developing countries face structural stumbling blocks in the health sector. They tend to have an underdeveloped health infrastructure with limited resources and a lack of health professionals (Tran et al., 2020). This situation is aggravated during the pandemic due to lack of protective gears for the health care personnel, lack of test kids and hospital facilities that make it difficult to separate COVID cases from the other patients (Giri & Rana, 2020). The risk of infection spread from COVID to non-COVID patients and health professionals is high. Consequently and similar to developed countries, health professionals in developing countries are severely affected by the pandemic resulting in an immediate aggravation of health sector struggles (Arthur-Holmes et al., 2020; Usman et al., 2020).

A careful analysis of the presented evidence about the potential indirect effects of COVID-19 on health care seeking, population health, and the health sector reveals that most estimates and evidence results from immediate responses and small (purposive) samples. The current challenge is to obtain reliable data across countries and health systems that allow to assess which of the alleged, possible large indirect health impacts have or will materialize and which ones have been overlooked in the early calculations. In an attempt to assess the costs of foregone treatment I present some exploratory calculations based on hospital admissions data in the next section.

8.3 Exploratory Analysis of the Economic Costs Associated with Delayed and Foregone Hospital Care

The costs associated with delayed and foregone hospital admissions is assessed employing data from Eurostat (2021) about hospital discharges. While ideally, I would have liked to assess data from developing countries, they are extremely scarce and rather outdated. Therefore, I start the analysis with the more detailed data from Eurostat and then extrapolate the findings in a second step for those developing countries where information on in-patient care use is available for any point in time during the last 10 years. For this second analysis I employ information from the Health Equity and Financial Protection Database (2021).

Eurostat provides the total number of hospital discharges of in-patients for all causes of disease (excluding accidents and births) for the period 2009 to 2018. In the analysis I focus on available data for Central and Eastern European countries. A detailed overview of the data along with the yearly growth rate in hospital discharges and the average growth rate over the period is provided in appendix A1. The dataset contains 8 countries of which all except one (Bulgaria) are classified as high-income countries. Clearly, the countries do not represent developing countries. Yet, many of them were members of the former Soviet Union and only recently became high income countries and present thus a good initial approximation since credible yearly data is available for these countries contrary to the developing countries.

Based on the average growth rate in hospital discharges I derive estimates of 2019 and 2020 hospital discharges not considering any impacts of COVID-19 as I am interested in operations as usual and impacts on these. Results are shown in Table 8.1, Columns 1 and 2.

According to US data for 162 hospitals spanning 21 states and covering 22 million patients, hospital admissions declined by 6.9% between March 8 and August 8, 2020 (Heist et al., 2020). This figure increase to 8.5% if all available data until December 5, 2020 are taken into account (Heist et al., 2021). Based on these data I estimate two scenarios, the first one assumes a 5% decline in regular hospital admissions unrelated to COVID-19 cases and the second one a 10% reduction. There is no doubt that delayed care is associated with increased costs and worse health outcomes (Kraft et al., 2009). Yet, the precise costs are hard to determine. For the sake of this exploration, I put potential costs in relation to GDP per capita. Imposing the assumption that delayed health care seeking increases medical costs, puts higher

8 Indirect Health Effects Due to COVID-19: An Exploration of Potential Economic...

Table 8.1 Economic costs due to a COVID-related decrease in regular hospital discharges, Central and Eastern European countries

Country	Annual hospital discharges (HD): Estimates 2019 (1)	2020 (2)	2019 GDP per capita (constant 2010 US$) (3)	2019 GDP (constant 2010 US$) (4)	3 scenarios: Total economic costs of a ... 5% decline in HD, 50% decline in yearly GDP pc (5)	10% decline in HD, 50% decline in yearly GDP pc (6)	10% decline in HD, 100% decline in yearly GDP pc (7)	3 scenarios: Reduction in GDP (%) 5% decline in HD, 50% decline in yearly GDP pc (8)	10% decline in HD, 50% decline in yearly GDP pc (9)	10% decline in HD, 100% decline in yearly GDP pc (10)
Bulgaria	2,456,771	2,513,044	9059	63,191,572,278	569,141,590	1,138,283,181	2,276,566,361	0.90	1.80	3.60
Czechia	2,071,523	2,060,718	24,266	258,911,043,396	1,250,134,766	2,500,269,531	5,000,539,062	0.48	0.97	1.93
Croatia	654,790	647,908	16,510	67,154,066,753	267,423,844	534,847,688	1,069,695,377	0.40	0.80	1.59
Hungary	1,864,182	1,846,284	17,572	171,680,578,831	811,072,690	1,622,145,379	3,244,290,759	0.47	0.94	1.89
Poland	7,060,193	7,586,745	17,407	660,941,995,821	3,301,561,926	6,603,123,852	13,206,247,703	0.50	1.00	2.00
Romania	3,998,031	3,885,852	12,092	234,057,305,195	1,174,693,157	2,349,386,314	4,698,772,628	0.50	1.00	2.01
Slovenia	363,807	364,782	27,427	57,265,665,047	250,121,731	500,243,461	1000,486,922	0.44	0.87	1.75
Slovakia	1,040,176	1,040,343	20,999	114,530,788,157	546,153,975	1,092,307,950	2,184,615,901	0.48	0.95	1.91
Average	2,438,684	2,493,210	18,167	203,466,626,935	1,021,287,960	2,042,575,920	4,085,151,839	0.52	1.04	2.08

Note: The annual hospital discharges for 2019 and 2020 represent estimates based on Eurostat data for the years 2009 to 2018 (compare Table 8.3 in the Appendix for details). GDP data are taken from the World Development Indicators (2021). HD abbreviates hospital discharges

demands on family care needs, and deteriorates long-term health and productivity outcomes, I impose that every delayed in-patient treatment independent of the cause results in a 50–100% reduction of annual GDP per capita. While this is a gross simplification and might seem like a large figure, it is meant to also indirectly account for reduced out-patient care seeking for which detailed data are not available. Rice et al. (1985) calculate the total economic costs of illness in 1980 at $455 billion for the United States. That amounts to 7% of the 1980s GDP (constant 2010 US$). In relation to these large costs, imposing that delayed and foregone medical care amounts to costs of 50–100% of annual GDP per capita is a decent approximation to get an initial idea of possible economic costs.

I present three scenarios. The first scenario assumes a 5% decline in regular hospital discharges (corresponds roughly to the short-term figures from the US, compare Heist et al., 2020) as a consequence of the pandemic and that every one of these untreated patients leads to an average economic cost that is equivalent to a decline of 50% in yearly GDP per capita. The total costs average more than one billion US$ across the eight countries under study, Slovenia has the lowest costs of around 250 million US$ and Poland has the highest costs of more than 3.3 billion US$ (Table 8.1, Column 5). In terms of reductions of overall GDP, the economic costs of delayed and foregone health care are expected to amount to about 0.5% of total GDP on average (Table 8.1, Column 8). The largest impact can be observed for Bulgaria. The costs of delayed and foregone health care seeking amount to 0.9% of GDP. The smallest relative impact is found in Croatia where the costs represent 0.4% of GDP. The second scenario imposes a 10% decrease in regular hospital discharges and is akin to the long-term figures from the United States (Heist et al., 2021). Unsurprisingly, this results in twice the costs and an average decrease in total GDP of roughly 1% (Table 8.1, Columns 6 and 9). If I further impose that the economic cost of each delayed and foregone hospital admission amounts to the full annual GDP per capita impacts again double (Table 8.1, Columns 7 and 10). While these scenarios are simplistic and the assumption of a 50–100% loss in yearly GDP per capita is strong, they give a first indication that the economic costs stemming from delayed and foregone health care seeking are not to be underestimated. Importantly, some of these costs will be incurred over many years to come due to long-term negative health consequences and chronic conditions.

The exercise of estimating the total economic costs of delayed and foregone health care is repeated for in-patient data from Sub-Sahara Africa obtained from the Health Equity and Financial Protection Database (2021). No patient numbers are available but population shares that had in-patient treatment in the reporting year. Particular caution has to be applied in interpreting these data since the challenge is that the available in-patient data are up to 10 years old. Older data have not been used resulting in only eleven countries being represented in the analysis. No temporal accounting was applied to the data since such exercise is subject to too much uncertainty. The in-patient shares, no matter which year they were from, were directly applied to the 2019 population data. I use 2019 population data since 2020 data is potentially affected by COVID-related death. The same three scenarios were calculated as for the Central and Eastern European countries. Results are presented in Table 8.2. As can be seen even at a glance the average economic costs per country are substantially lower for the Sub-Saharan African countries under study when imposing that they face the same 5–10% reduction in hospital discharges from regular diseases as developed countries. The resulting economic costs range between 74 million US$ and 297 million US$ on average (Table 8.2). These lower average costs represent a mechanic result since costs have been calculated relative to the countries' GDP and the Sub-Saharan African countries under study have considerably lower GDPs per capita compared to the Central and Eastern European countries under study. Yet, also in terms of the share of these indirect health costs in GDP the Sub-Saharan African countries face a lower burden ranging between 0.1% and 0.6% of total GDP on average. This reflects the considerably lower use of in-patient care in Sub-Sahara Africa. In the dataset at hand it amounts to 5.7% of the total population on average (Table 8.2). The share of hospital discharges for the Central and Eastern European Countries under study is almost 21%. While these figures cannot necessarily be directly compared, since they are the result of a different underlying accounting metric, they give a clear indication of structural differences, i.e. the pre-pandemic lack of available health infrastructure in many Sub-Sahara African countries. Thus, the resulting cost figures are smaller by construction. If, however, I assume a 50% decrease in "normal" in-patient care, which is not unrealistic given the hurdles for accessing care in Sub-Sahara Africa and the fear the pandemic caused because of knowledge of the impacts of other pandemics such as Ebola (Wadoum and Clarke, 2020), then I obtain estimates that are similar to the relative loss in GDP in the Central and Eastern European countries. Imposing a 50% decline in hospital admissions and a 50% (100%) reduction in GDP per capita leads to a 1.4% (almost 3%) decline in overall GDP on average. If the costs are in the range of 1–3% of GDP they counteract the already low GDP growth in Sub-Sahara Africa and have the potential to leave the African continent further behind.

As these calculations and considerations show, it is not possible to fully gauge the economic costs of delayed and foregone health care due to COVID-19 in developing countries. Yet, the presented rough calculations in combination with the modelling results about excess mortality caution against exclusively focusing our attention on fighting the spread of COVID-19 in developing countries without providing resources for the fight against other (endemic) diseases. Ignoring the burden of currently untreated health conditions including the mental distress caused by the pandemic potentially makes us to severely underestimate the long-term effects of the pandemic for the populations of developing countries and for achieving more equity and less poverty in the years to come. Moreover, the available in-patient data for Sub-Sahara Africa leaves no doubt about the need to further expand health services in the region not only for the fight of the current pandemic but also to sustainably increase population health and to permanently reduce mortality from untreated conditions. Last but not least the lack of quality data calls for a strengthening of health information systems.

Table 8.2 Economic costs due to a COVID-related decrease in regular hospital discharges, Sub-Sahara African countries

Country	Year of most recent in-patient data	Annual in-patient care use (share of population 18+)	Population (2019)	Population based estimate of in-patient care use	2019 GDP per capita (constant 2010 US$)	2019 GDP (constant 2010 US$)	3 scenarios: Total economic costs of a ... 5% decline in HD, 50% decline in yearly GDP pc	10% decline in HD, 50% decline in yearly GDP pc	10% decline in HD, 100% decline in yearly GDP pc	3 scenarios: Reduction in GDP (%) 5% decline in HD, 50% decline in yearly GDP pc	10% decline in HD, 50% decline in yearly GDP pc	10% decline in HD, 100% decline in yearly GDP pc
Congo, Dem. Rep.	2012	5.94%	86,790,567	5,159,237	424	36,767,977,989	54,641,517	109,283,034	218,566,069	0.15	0.30	0.59
Ethiopia	2015	2.52%	112,078,730	2,825,420	603	67,542,461,143	42,567,353	85,134,706	170,269,411	0.06	0.13	0.25
Gambia	2015	1.33%	2,347,706	31,246	815	1,913,886,427	636,799	1,273,598	2,547,195	0.03	0.07	0.13
Ghana	2012	10.20%	30,417,856	3,103,747	1884	57,315,905,531	146,208,576	292,417,153	584,834,305	0.26	0.51	1.02
Kenya	2015	4.22%	52,573,973	2,217,781	1237	65,060,160,105	68,612,461	137,224,922	274,449,844	0.11	0.21	0.42
Liberia	2014	8.13%	4,937,374	401,519	516	2,548,948,907	5,182,165	10,364,330	20,728,659	0.20	0.41	0.81
Malawi	2016	4.20%	18,628,747	782,868	524	9,754,076,228	10,247,813	20,495,626	40,991,252	0.11	0.21	0.42
Niger	2011	4.17%	23,310,715	972,871	563	13,127,408,385	13,696,790	27,393,581	54,787,162	0.10	0.21	0.42
Nigeria	2012	3.15%	200,963,599	6,335,257	2374	477,161,826,016	376,056,021	752,112,041	1,504,224,082	0.08	0.16	0.32
Sierra Leone	2011	12.79%	7,813,215	999,492	488	3,816,410,758	12,205,192	24,410,383	48,820,767	0.32	0.64	1.28
Tanzania	2015	5.98%	58,005,463	3,468,343	985	55,482,146,264	85,446,799	170,893,598	341,787,197	0.15	0.31	0.62
Average		5.70%	54,351,631	2,390,707	947	71,862,837,068	74,136,499	148,272,997	296,545,995	0.14	0.29	0.57

Note: Annual in-patient care data stems from the Health Equity and Financial Protection Database, only information about Sub-Sahara African countries not older than 10 years is considered. Population and GDP data are taken from the World Development Indicators

8.4 Conclusions

Developing countries face a trade-off when it comes to the current pandemic. On the one hand, demography, i.e. the relative youth of the population of most developing countries, is likely to play in their favor since most of the severe COVID-19 cases and deaths happen among the elderly. On the other hand, the indirect health consequences are likely to hit these countries even harder due to poor health systems as well as delayed and foregone health care and the long-term negative implications delayed care has on productivity. Although difficult to quantify, the indirect health effects loom large since they do not only materialize in the economic cost of delayed care but also in pre-mature productivity losses due to additional deaths from disruptions in the care of HIV, tuberculosis and malaria patients to name just the main diseases.

Financial support that goes beyond the provision of COVID-19 test kits and vaccines needs to be provided to ensure that the Africa region can continue its fight against the endemic diseases that pre-date COVID-19 and that the already low levels of growth can at least be maintained. As the exploratory economic cost calculations have shown, the indirect health effects due to untreated diseases have the potential to leave Africa behind if not attended to. In addition, there is the new burden from fear, anxiety and mental diseases among health care worker in particular and the larger population in general. Developed countries do good in supporting developing countries in their health system responses to COVID-19 even if the main objective of the former is to avoid further Coronavirus mutations and their spread across the globe. Leaving developing countries behind now, has the potential to not only magnify the direct and indirect health costs experienced in developing countries but also the already extremely high direct and indirect health costs in developed countries.

Two practical and short-term solutions to help developing countries cope are provided by (i) self-tests and (ii) mobile health interventions. Self-tests, not only for COVID-19 but also for HIV and TB (Mhango et al., 2020; Xie et al., 2012), can act as one puzzle to curb direct as well as indirect health effects of the pandemic. In addition, the role for mobile health interventions and tele-medicine should not be underestimated and further supported across developing countries (Kadir, 2020; Keri et al., 2020), at least for follow-up consultations (Kumar et al., 2020). Pragmatic and hands-on solutions are needed to strengthen the health systems of developing countries (Gerard et al., 2020). To facilitate entry, interventions can build on existing structures (decentralized health services, community insurance systems, (un) conditional cash transfers) and involve locally respected experts (local and religious leaders, NGO trained community volunteers, traditional healers).

Appendix

Table 8.3 Yearly hospital discharges, in-patients, total number and associated yearly growth rates

Yearly hospital discharges, in-patients, total number: Existing Data

Country	Country income classification	2009	2010	2011	2012	2013	2014	2015	2016	2017	2018	Average growth rate
Bulgaria	U-MI	1,958,897	1,917,199 (−0.021)	1,961,177 (0.023)	2,040,666 (0.041)	2,221,115 (0.088)	2,323,313 (0.046)	2,302,891 (−0.009)	2,258,579 (−0.019)	2,331,264 (0.032)	2,401,759 (0.030)	0.023
Czechia	HI	2,182,739	2,156,906 (−0.012)	2,119,827 (−0.017)	2,107,882 (−0.006)	2,142,523 (0.016)	2,171,104 (0.013)	2,145,844 (−0.012)	2,112,698 (−0.015)	2,099,139 (−0.006)	2,082,385 (−0.008)	−0.005
Croatia	HI	727,763	683,435 (−0.061)	692,418 (0.013)	664,139 (−0.041)	669,855 (0.009)	664,564 (−0.008)	674,820 (0.015)	674,893 (0.000)	661,716 (−0.020)	661,745 (0.000)	−0.011
Hungary	HI	2,123,781	2,052,238 (−0.034)	2,045,775 (−0.003)	1,995,583 (−0.025)	1,998,524 (0.001)	2,007,115 (0.004)	1,969,389 (−0.019)		1,903,738 (−0.011)	1,882,253 (−0.011)	−0.010
Poland	HI	3,438,959	3,411,640 (−0.008)	3,432,659 (0.006)	6,251,264 (0.821)	6,352,351 (0.016)	6,513,243 (0.025)	6,304,164 (−0.032)	6,805,517 (0.080)	6,893,122 (0.013)	6,570,185 (−0.047)	0.075
Romania	HI	5,314,291	5,048,421 (−0.050)	4,633,328 (−0.082)	4,431,233 (−0.044)	4,449,658 (0.004)	4,234,437 (−0.048)	4,146,344 (−0.021)	4,069,680 (−0.018)	4,068,326 (0.000)	4,113,449 (0.011)	−0.028
Slovenia	HI	354,197	350,966 (−0.009)	358,163 (0.021)	351,914 (−0.017)	373,899 (0.062)	379,517 (0.015)	380,862 (0.004)	377,043 (−0.010)	363,952 (−0.035)	362,834 (−0.003)	0.003
Slovakia	HI	1,015,428	1,012,831 (−0.003)	991,583 (−0.021)	1,029,301 (0.038)		1,052,702 (0.015)	1,061,730 (0.009)	1,071,654 (0.009)	1,061,067 (−0.010)	1,040,010 (−0.020)	0.000

Note: The data is taken from Eurostat (2021). Yearly growth rates in parentheses. The average growth rate across years is calculated as geometric mean. In the country income classification U-MI abbreviates upper-middle income country and HI high-income country

References

Arthur-Holmes, F., Akaadom, M. K. A., Agyemang-Duah, W., Busia, K. A., & Peprah, P. (2020). Healthcare concerns of older adults during the COVID-19 outbreak in low- and middle-income countries: Lessons for health policy and social work. *Journal of Gerontological Social Work, 63*(6–7), 717–723. https://doi.org/10.1080/01634372.2020.1800883

Del Pilar Estevez-Diz, M., Bonadio, R. C., Miranda, V. C., & Carvalho, J. P. (2020). Management of cervical cancer patients during the COVID-19 pandemic: a challenge for developing countries. *Ecancermedicalscience, 14*, 1060. https://doi.org/10.3332/ecancer.2020.1060

Dhandapani, M., & Dhandapani, S. (2020). Challenges posed by COVID-19 and neurosurgical nursing strategies in developing countries. *Surgical Neurology International, 11*. https://doi.org/10.25259/SNI_677_2020

Drissi, N., Ouhbi, S., Marques, G., de la Torre Díez, I., Ghogho, M., & Idrissi, M. A. J. (2020). A systematic literature review on e-mental health solutions to assist health care workers during COVID-19. *Telemedicine and e-Health*. https://doi.org/10.1089/tmj.2020.0287

Einstein, A. J., Shaw, L. J., Hirschfeld, C., Williams, M. C., Villines, T. C., Better, N., Vitola, J. V., Cerci, R., Dorbala, S., Raggi, P., Choi, A. D., Lu, B., Sinitsyn, V., Sergienko, V., Kudo, T., Nørgaard, B. L., Maurovich-Horvat, P., Campisi, R., Milan, E., et al. (2021). International impact of COVID-19 on the diagnosis of heart disease. *Journal of the American College of Cardiology, 77*(2), 173–185.

Eurostat. (2021). https://appsso.eurostat.ec.europa.eu/nui/submitViewTableAction.do. Last accessed: 15 Mar 2021.

Gerard, F., Imbert, C., & Orkin, K. (2020). Social protection response to the COVID-19 crisis: options for developing countries. *Oxford Review of Economic Policy, 36*(S1), S281–S296. https://doi.org/10.1093/oxrep/graa026

Giri, A. K., & Rana, D. (2020). Charting the challenges behind the testing of COVID-19 in developing countries: Nepal as a case study. *Biosafety and Health, 2*(2), 53–56. https://doi.org/10.1016/j.bsheal.2020.05.002

González-Montero, J., Valenzuela, G., Ahumada, M., Barajas, O., & Villanueva, L. (2020). Management of cancer patients during COVID-19 pandemic at developing countries. *World Journal of Clinical Cases, 8*(16), 3390–3404. https://doi.org/10.12998/wjcc.v8.i16.3390

Gupta, N., Goel, K., Shah, P., & Misra, A. (2012). Childhood obesity in developing countries: Epidemiology, determinants, and prevention. *Endocrine Reviews, 33*(1), 48–70.

Gupta, A. K., Mehra, A., Niraula, A., Kafle, K., Deo, S. P., Singh, B., Sahoo, S., & Grover, S. (2020). Prevalence of anxiety and depression among the healthcare workers in Nepal during the COVID-19 pandemic. *Asian Journal of Psychiatry, 54*, 102260. https://doi.org/10.1016/j.ajp.2020.102260

Haider, I. I., Tiwana, F., & Tahir, S. M. (2020). Impact of the COVID-19 pandemic on adult mental health. *Pakistan Journal of Medical Sciences, 36*(COVID19-S4), S90–S94. https://doi.org/10.12669/pjms.36.COVID19-S4.2756

Health Equity and Financial Protection Database. (2021). https://databank.worldbank.org/source/health-equity-and-financial-protection-indicators-(hefpi). Last accessed: 11 Mar 2021.

Heist, T., Schwartz, K., & Butler, S. (2020). *How were hospital admissions impacted by COVID-19? Trends in overall and non-COVID-19 hospital admissions through august 8, 2020.* https://www.kff.org/health-costs/issue-brief/how-were-hospital-admissions-impacted-by-covid-19-trends-in-overall-and-non-covid-19-hospital-admissions-through-august-8-2020/. Last accessed: 9 Mar 2021.

Heist, T., Schwartz, K.., & Butler, S. (2021). *Trends in overall and non-COVID-19 hospital admissions.* Issue Brief. https://www.kff.org/health-costs/issue-brief/trends-in-overall-and-non-covid-19-hospital-admissions/. Last accessed: 9 Mar 2021.

Hussein, M. I. H., Albashir, A. A. D., Elawad, O. A. M. A., & Homeida, A. (2020). Malaria and COVID-19: unmasking their ties. *Malaria Journal, 19*, 457. https://doi.org/10.1186/s12936-020-03541-w

Jain, V. K., Iyengar, K. P., Samy, D. A., & Vaishya, R. (2020). Tuberculosis in the era of COVID-19 in India. *Diabetes & Metabolic Syndrome: Clinical Research & Reviews, 14*(5), 1439–1443. https://doi.org/10.1016/j.dsx.2020.07.034

Jewell, B. L., Mudimu, E., Stover, J., Ten Brink, D., Phillips, A. N., Smith, J. A., Martin-Hughes, R., Teng, Y., Glaubius, R., Mahiane, S. G., Bansi-Matharu, L., Taramusi, I., Chagoma, N., Morrison, M., Doherty, M., Marsh, K., Bershteyn, A., Hallett, T. B., Kelly, S. L., & the HIV Modelling Consortium. (2020). Potential effects of disruption to HIV programmes in sub-Saharan Africa caused by COVID-19: results from multiple mathematical models. *The Lancet HIV, 7*(9), e629–e640. https://doi.org/10.1007/s10461-020-02935-w

Kadir, M. A. (2020). Role of telemedicine in healthcare during COVID-19 pandemic in developing countries. *Telehealth and Medicine Today, 5*(2). https://doi.org/10.30953/tmt.v5.187

Keri, V. C., Brunda, R. L., Sinha, T. P., Wig, N., & Bhoi, S. (2020). Tele-healthcare to combat COVID-19 pandemic in developing countries: A proposed single centre and integrated national level model. *The International Journal of Health Planning and Management, 35*(6), 1617–1619. https://doi.org/10.1002/hpm.3036

Koh Boon Yau, E., Ping, N. P. T., Shoesmith, W. D., James, S., Hadi, N. M. N., & Loo, J. L. (2020). The behaviour changes in response to COVID-19 pandemic within Malaysia. *The Malaysian Journal of Medical Sciences, 27*(2), 45–50. https://doi.org/10.21315/mjms2020.27.2.5

Kraft, A. D., Quimbo, S. A., Solon, O., Shimkhada, R., Florentino, J., & Peabody, J. W. (2009). The health and cost impact of care delay and the experimental impact of insurance on reducing delays. *Journal of Pediatrics, 155*(2), 281-5.e1. https://doi.org/10.1016/j.jpeds.2009.02.035

Kumar, S., Bishnoi, A., & Vinay, K. (2020). Changing paradigms of dermatology practice in developing nations in the shadow of COVID-19: Lessons learnt from the pandemic. *Dermatologic Therapy, 33*(4), e13472. https://doi.org/10.1111/dth.13472

Lagat, H., Sharma, M., Kariithi, E., Otieno, G., Katz, D., Masyuko, S., Mugambi, M., Wamuti, B., Weiner, B., & Farquhar, C. (2020). Impact of the COVID-19 pandemic on HIV testing and assisted partner notification services, Western Kenya. *AIDS and Behavior, 24*, 3010–3013.

Mai, T. M. T., Pham, N. O., Tran, T. M. H., Baker, P., Gallegos, D., Do, T. N. D., van der Pols, J. C., & Jordan, S. J. (2020). The double burden of malnutrition in Vietnamese school-aged children and adolescents: a rapid shift over a decade in Ho Chi Minh City. *European Journal of Clinical Nutrition, 74*, 1448–1456. https://doi.org/10.1038/s41430-020-0587-6

Mamun, M. A., Bodrud-Doza, M., & Griffiths, M. D. (2020). Hospital suicide due to non-treatment by healthcare staff fearing COVID-19 infection in Bangladesh? *Asian Journal of Psychiatry, 54*, 102295. https://doi.org/10.1016/j.ajp.2020.102295

Mhango, M., Chitungo, I., & Dzinamarira, T. (2020). COVID-19 lockdowns: impact on facility-based HIV testing and the case for the scaling up of Home-based testing services in Sub-Saharan Africa. *AIDS Behavior, 24*(11), 3014–3016. https://doi.org/10.1007/s10461-020-02939-6

Nouhjah, S., & Jahanfar, S. (2020). Challenges of diabetes care management in developing countries with a high incidence of COVID-19: A brief report. *Diabetes & Metabolic Syndrome: Clinical Research & Reviews, 14*(5), 731–732. https://doi.org/10.1016/j.dsx.2020.05.012

O'Leary, A., Jalloh, M. F., & Neria, Y. (2018). Fear and culture: Contextualising mental health impact of the 2014–2016 Ebola epidemic in West Africa. *BMJ Global Health, 3*, e000924. https://doi.org/10.1136/bmjgh-2018-000924

Pifarré i Arolas, H. E., Acosta, E., López-Casasnovas, G., Lo, A., Nicodemo, C., Riffe, T., & Myrskylä, M. (2021). Years of life lost to COVID-19 in 81 countries. *Scienctific Reports, 11*(3504). https://doi.org/10.1038/s41598-021-83040-3

Ponticiello, M., Mwanga-Amumpaire, J., Tushemereirwe, P., Nuwagaba, G., King, R., & Sundararajan, R. (2020). "Everything is a Mess": How COVID-19 is impacting engagement with HIV testing services in rural Southwestern Uganda. *AIDS and Behavior, 24*, 3006–3009.

Popkin, B. M., Adair, L. S., & Ng, S. W. (2012). Global nutrition transition and the pandemic of obesity in developing countries. *Nutrition Reviews, 70*(1), 3–21. https://doi.org/10.1111/j.1753-4887.2011.00456.x.

Poudel, K., & Subedi, P. (2020). Impact of COVID-19 pandemic on socioeconomic and mental health aspects in Nepal. *International Journal of Social Psychiatry, 66*(8), 748–755. https://doi.org/10.1177/0020764020942247

Rice, D. P., Hodgson, T. A., & Kopstein, A. N. (1985). The economic costs of illness: A replication and update. *Health Care Financing Review, 7*(1), 61–80.

Roberton, T., Carter, E. D., Chou, V. B., Stegmuller, A. R., Jackson, B. D., Tam, Y., Sawadogo-Lewis, T., & Walker, N. (2020). Early estimates of the indirect effects of the COVID-19 pandemic on maternal and child mortality in low-income and middle-income countries: A modelling study. *The Lancet Global Health, 8*, e901–e908. https://doi.org/10.1016/S2214-109X(20)30229-1

Sarin, Y. K. (2020). Resuming elective surgeries in Corona pandemic from the perspective of a developing country. *Journal of Pediatric and Adolescent Surgery, 1*(1), 19–21. https://doi.org/10.46831/jpas.v1i1.27

Saunders, M. J., & Evans, C. A. (2020). COVID-19, tuberculosis and poverty: Preventing a perfect storm. *European Respiratory Journal, 56*, 2001348. https://doi.org/10.1183/13993003.01348-2020

Tran, B. X., Vu, G. T., Le, H. T., Pham, H. Q., Phan, H. T., Latkin, C. A., & Ho, R. C. (2020). Understanding health seeking behaviors to inform COVID-19 surveillance and detection in resource-scarce settings. *Journal of Global Health, 10*(2), 0203106. https://doi.org/10.7189/jogh.10.0203106

UNICEF. (2020). *Pregnant mothers and babies born during COVID-19 pandemic threatened by strained health systems and disruptions in services.* https://www.unicef.org/press-releases/pregnant-mothers-and-babies-born-during-covid-19-pandemic-threatened-strained-health. Last accessed: 9 Mar 2021.

UNICEF. (2021). *COVID-19 A threat to progress against child marriage.* UNICEF.

Usman, N., Mamun, M. A., & Ullah, I. (2020). COVID-19 infection risk in pakistani healthcare workers: The cost-effective safety measures for developing countries. *Social Health and Behaviour, 3*(3), 75–77.

Varshney, M., Parel, J. T., Raizada, N., & Sarin, S. K. (2020). Initial psychological impact of COVID-19 and its correlates in Indian Community: An online (FEEL-COVID) survey. *PLoS One, 15*(5), e0233874. https://doi.org/10.1371/journal.pone.0233874

Wadoum, R. E. G., & Clarke, A. (2020). How prepared is Africa to face COVID-19? *The Pan African Medical Journal, 35*(2), 1. https://doi.org/10.11604/pamj.supp.2020.35.2.22665

Walker, P. G. T., White, M. T., Griffin, J. T., Reynolds, A., Ferguson, N. M., & Ghani, A. C. (2015). Malaria morbidity and mortality in Ebola-affected countries caused by decreased healthcare capacity, and the potential effect of mitigation strategies: a modelling analysis. *Lancet Infectious Diseases, 15*(7), 825–832. https://doi.org/10.1016/S1473-3099(15)70124-6

Wang, C., Pan, R., Wan, X., Tan, Y., Xu, L., Ho, C. S., & Ho, R. C. (2020). Immediate psychological responses and associated factors during the initial stage of the 2019 coronavirus disease (COVID-19) epidemic among the general population in China. *International Journal of Environmental Research and Public Health, 17*, 1729. https://doi.org/10.3390/ijerph17051729

WHO. (2020a). *Access to HIV medicines severely impacted by COVID-19 as AIDS response stalls.* News release. https://www.who.int/news/item/06-07-2020-who-access-to-hiv-medicines-severely-impacted-by-covid-19-as-aids-response-stalls. Last accessed: 9 Mar 2021.

WHO. (2020b). *The cost of inaction: COVID-19-related service disruptions could cause hundreds of thousands of extra deaths from HIV.* News release. https://www.who.int/news/item/11-05-2020-the-cost-of-inaction-covid-19-related-service-disruptions-could-cause-hundreds-of-thousands-of-extra-deaths-from-hiv. Last accessed: 10 Mar 2021.

WHO. (2020c). *World TB Day.* Fact Sheet. https://www.afro.who.int/health-topics/tuberculosis-tb. Last accessed: 11 Mar 2021.

WHO. (2020d). *Global tuberculosis report 2020.* World Health Organization.

WHO. (2020e). *Malaria.* Fact Sheet. https://www.who.int/news-room/fact-sheets/detail/malaria. Last accessed: 11 Mar 2021.

WHO. (2020f). *The potential impact of health service disruptions on the burden of malaria: A modelling analysis for countries in sub-Saharan Africa*. World Health Organization.

World Development Indicators. (2021). https://databank.worldbank.org/source/world-development-indicators. Last accessed: 11 Mar 2021.

Xie, H., Mire, J., Kong, Y., Chang, M., Hassounah, H. A., Thornton, C. N., Sacchettini, J. C., Cirillo, J. D., & Rao, J. (2012). Rapid point-of-care detection of the tuberculosis pathogen using a BlaC-specific fluorogenic probe. *Nature Chemistry, 4*, 802–809. https://doi.org/10.1038/nchem.1435

Xie, X., Xue, Q., Zhou, Y., Zhu, K., Liu, Q., Zhang, J., & Song, R. (2020). Mental Health Status Among Children in Home Confinement During the Coronavirus Disease 2019 Outbreak in Hubei Province, China. *JAMA Pediatrics, 174*(9), 898–900. https://doi.org/10.1001/jamapediatrics.2020.1619

Yang, L., Yin, J., Wang, D., Rahman, A., & Li, X. (2020). Urgent need to develop evidence-based self-help interventions formental health of healthcare workers in COVID-19 pandemic. *Psychological Medicine*, 1–2. https://doi.org/10.1017/S0033291720001385

Zhang, W., Wang, K., Yin, L., Zhao, W., Xue, Q., Peng, M., Min, B., Tian, Q., Leng, H., Du, J., Chang, H., Yang, Y., Li, W., Shangguan, F., Yan, T., Dong, H., Han, Y., Wang, Y., Cosci, F., & Wang, H. (2020). Mental health and psychosocial problems of medical health workers during the COVID-19 epidemic in China. *Psychotherapy and Psychosomatics, 89*, 242–250. https://doi.org/10.1159/000507639

Chapter 9
Effects of COVID-19 on Education and Schools' Reopening in Latin America

Georgina M. Gómez and G. J. Andrés Uzín P.

Abstract Lockdowns around the world have forced governments to close schools for most of 2020 and they will probably remain closed in several areas and for some time until the end of 2021. Millions of children around the world will not have experienced classes in classrooms for an entire year or longer, which has disrupted their routines and educational cycles. In this chapter we discuss the impact of the pandemic and school closures on the schooling system in Latin America. A rapid shift to digital learning became necessary, but the basic infrastructure to support it was mostly absent, so school closures have exacerbated the digital gap both within and between countries. In Latin America, after decades of expansion of the public schooling system, the pandemic has offset the important progress in terms of poverty alleviation.

9.1 Introduction

The effects of the Covid-19 pandemic on the education sector will be deeper and will last longer than we can envision at the time of writing this chapter. Social distancing and other non-pharmaceutical measures have persuaded many countries to close schools, universities, and other teaching centres in an effort to protect the health of the population. Closures of learning centres have made it impossible for learners and teachers to meet in classrooms and this distancing has seriously disrupted their learning processes and their life-time employment opportunities.

G. M. Gómez (✉)
International Institute of Social Studies of Erasmus University Rotterdam, Rotterdam, The Netherlands
e-mail: gomez@iss.nl

G. J. Andrés Uzín P.
Graduate School of Public Management and Olave School of Business, Universidad Privada de Bolivia (UPB), La Paz, Bolivia

© The Author(s), under exclusive license to Springer Nature Switzerland AG 2022
E. Papyrakis (ed.), *COVID-19 and International Development*,
https://doi.org/10.1007/978-3-030-82339-9_9

The consequences of closing learning centres can be summarized along several lines. First, the lockdowns driven by the need to protect public health have disrupted the learning processes of learners. Research in the US has shown that even routine interruptions such as summer vacations both deactivates learning capacities and causes the loss of acquired knowledge (Alexander et al., 2007; Cooper, 1996). Interruptions also have an impact on absenteeism and school desertion, which may be temporary or permanent. In the medium and long run, the restrictions on traditional methods of schooling will translate in reduced employment and social opportunities for the youth affected.

While some course contents will be delivered with the use of new technologies, these are far from given. The second impact of school closures has been termed digital gap or digital divide, and it addresses the unequal access to new information and communication technologies as well as the differentiation in the abilities to leverage on technologies for socio-economic development (Mariscal et al., 2019). In the last decade, the specialized literature acknowledged the widening of the digital gap among income groups, but the pandemic has significantly exacerbated these inequalities within and between countries and its consequences on education.

Learning centres are not simply large buildings with classrooms that children attend to acquire knowledge. Schools generate learning communities that involves households, neighbours and a wide range of actors for whom the learning centre is a meeting point. The third broad consequence of the lockdowns will impact a generation of children and youth that have missed on experiencing the role of the school in promoting social cohesion and connecting communities, even if students manage to achieve some of the learning objectives in their courses' contents.

In a nutshell, the pandemic has already marked the schooling of a generation that missed out learning contents and educational experiences. This chapter will first discuss the impact of school closures caused by the Covid-19 pandemic in Latin America. We have chosen to focus on Latin America because in the last two decades most of the countries in the region have given priority in its policies and investments to promote education as means to offset the severe historical inequalities in Latin American societies. At the same time, it has been one of the continents worst affected by the pandemic in terms of numbers of infections, deaths, and days with lockdowns. As a result, the impact of the pandemic and lockdown policies on schools have been unprecedented across the region. In some countries, schools have been closed for a whole year and in others, new school closures have not been discarded (Seusan & Sachs-Israel, 2020). The situation continues to unfold and further peaks of infections can be expected in 2021 and 2022.

This chapter is structured as follows. We first introduce the evolution of the education sector in Latin America in the last decades. We then describe the economic implications of the lockdown and the implications of school closures during the pandemic, the missed learning content, and the disruption to the lives of local communities centred on schools. We subsequently analyse the digital gap as it has widened, and we finally take stock of the challenges that lie ahead to reopen the education systems.

9.2 Promotion of Public Education

Latin American countries have made significant progress in expanding and improving education systems in the last five decades. To better understand the tragic setback caused by the Covid-19 pandemic, this section will briefly account for the development of the public education sector, especially in relation to school coverage.

Since the 1960s, with the consolidation of the public sector in Latin American countries, education was placed as a relatively high priority. While there are important variations among the countries in the region, in the last decades the emphasis was on expanding the public school system (Tedesco, 2012; Torres, 2005). The policy objective translated into two key goals: the provision of universal coverage to all children of school age and the harmonisation of the curricular content including schools of different ownership and modalities. Some governments had been relatively hands-off from the schooling system until then and basic or primary education was often provided by private actors, local communities, and religious groups. As national states consolidated and expanded their capacities, most governments multiplied the number of public schools across their national territories and the education systems moved gradually to a mix of public, semi-private and private schools. This mix, in turn, included various combinations of public and private ownership with public and private service delivery. Around the 1990s, several countries included decentralisation of the public schools' administration with central curricular development and a combination of sources of funding at the national and local levels, hence achieving a certain level of institutional sophistication.

The strategies pursued across Latin America, with variations among countries, included the increase of mandatory schooling time, expansion of primary school coverage to relatively neglected areas and social groups, and the creation of adjusted educational services for the marginalised, disabled and children with learning disorders or special needs. The priority was set on primary education, but the goals gradually extended to pre-schooling and secondary education. Table 9.1 shows the rates of coverage of primary and secondary schooling for six countries. As shown in Table 9.1, primary school coverage was already high at the turn of the Millennium and it was mostly perceived as an achieved goal. Secondary school rates were considerably lower but they were growing. Education policies in Latin America were certainly diverse, but a central concern was the reduction of inequalities via facilitating access to compulsory public schooling.

The 1990s presented additional challenges, with fiscal crises, budget cuts and neoliberal policies that froze investments in the education sector and compromised the efforts to expand the universal public schooling. Once these fiscal emergencies diminished, the priority on education resurfaced in most countries and since 2000 there has been a significant increase of the investment in education of all levels across the region. The rise is evident in absolute terms as well as in percentage of the GDP, as shown in Table 9.2 for the same group of countries. The target was to spend 4% of the GDP in public education, which became legally mandatory in some countries and a social demand, to the point that this level of expense figured

Table 9.1 Net schooling coverage, selected countries 1995–2015

Country	Bolivia		Colombia		Guatemala		Perú		Dominican Republic		Venezuela	
Year	Pri	Sec	Pri	Sec	Pri	Sec	Pri	Sec	Pri	Sec	Pri	Sec
1995	nd	nd	84.5	nd	nd	nd	84.9	52.2	nd	nd	87.3	23.5
2000	95.1	62.4	93.9	nd	82.4	24.5	96.8	62.0	84.7	40.6	95.4	25.9
2005	95.0	74.7	94.1	63.6	90.8	nd	96.6	67.2	84.3	53.3	93.0	68.2
2010	90.9	72.8	91.9	75.3	95.2	41.8	96.6	77.7	91.9	63.7	94.1	74.1
2015	87.3	76.7	91.9	75.6	85.2	44.1	92.7	79.2	93.5	67.8	93.0	73.2

Source: ECLAC (2020)
https://estadisticas.cepal.org/cepalstat/Perfil_Nacional_Social.html?pais=BOL&idioma=spanish
https://estadisticas.cepal.org/cepalstat/Perfil_Nacional_Social.html?pais=COL&idioma=spanish
https://estadisticas.cepal.org/cepalstat/Perfil_Nacional_Social.html?pais=GTM&idioma=spanish
https://estadisticas.cepal.org/cepalstat/Perfil_Nacional_Social.html?pais=PER&idioma=spanish
https://estadisticas.cepal.org/cepalstat/Perfil_Nacional_Social.html?pais=DOM&idioma=spanish
Note: net schooling rates report the percentage of children who are at school and should be at school at that age

Table 9.2 Public Education budget, as % of the GDP, selected countries

Country	Education Budget as % GDP
Bolivia	8.9
Colombia	4.0
Dominican Republic	4.0
Guatemala	2.8
Perú	3.5
Venezuela	4.9

Source: (UNESCO, 2015)

prominently in the 2012 winning presidential campaign of the Dominican Republic, for example.[1] As a reference, European countries spent an average of 5.6% of their GDP in education in 2010 (UNESCO, 2015).

The renewed efforts of the last two decades to achieve universal school coverage added to other policy priorities, which also varied by country but mostly included: a) the development of indicators to measure schooling quality, b) professional and economic improvement of the teaching staff, and c) adaptation of learning methods in changing contexts and implementation of new information technologies. These three challenges to improve the quality and modernise the education systems co-existed with the old pending goal of universal primary school coverage to finally include all children of school age in the primary or basic schools. The neglected social groups across Latin America that were still excluded from the public schooling system involved indigenous groups, remote rural areas, and urban areas severely affected by poverty and criminality (see also Arsel & Pellegrini, 2021).

---

[1] The social demand became known as "Movement of the 4%", see for example https://edujesuit.org/es/el-movimiento-de-reivindicacion-del-4-en-republica-dominicana/

The progress in the new educational agenda in Latin America was barely beginning to be assessed when economic growth started to slow down and education budgets were again frozen or reduced, as a result of the fall in international commodity prices. Since around 2015 several countries were again affected by public budget deficits, such as Colombia, Peru, Mexico, Chile and Argentina, in addition to the collapse of Venezuela. In these countries, investments and modernisation of the public schooling systems were slowing down and the third goal, the incorporation of new information technologies to digital learning, was progressing slowly or was repeatedly postponed (Tedesco, 2012; Lorente Rodríguez, 2019). In this context of relative budgetary weakness, the Covid-19 pandemic started and school closures were decided in most countries.

9.3 The Spread of Covid-19

In December, 2019 the SARS-CoV-2 new virus was discovered in the city of Wuhan, China, and isolated as the cause of the Covid-19 sickness that spread globally. The World Health Organization declared it was a public health emergency of International concern on January 30th, 2020 (WHO, 2020a) and later elevated the concern to a global pandemic on 11 March, 2020 (WHO, 2020b).

Although partial contingency plans existed in some countries, it was an unprecedented situation for which governments were badly prepared. The World Health Organization, with other multilateral organizations, designed and recommended a number of strategies through its Situation Reports (WHO, 2020a, c) prepared by its experts. Governments adopted some of these interventions to varying extents and with varying degrees of success, while a few governments completely ignored these recommendations. In general, policies to control the sickness and mitigate the effects of Covid-19 continue to generate considerable confusion and controversy.

Across the Americas the disease has not been adequately controlled and there have been with cycles of rise and decline in the numbers of infections and deaths. During the peaks, health care systems collapsed and the number of deaths in Latin America in the last year have been high. The region is home to 8.4% of the world population but of 18.6% of those globally tested positive for Covid-19. The data probably underestimates the real numbers of infections and deaths, considering the limited monitoring and testing facilities in the region. There is considerable variation among countries but 1 year after the beginning of the pandemic, the situation is still uncontrolled. On 25th March 2021, Brazil and Mexico registered over 320.000 and 200.000 deaths of Covid-19 patients respectively, with ratios of 153.49 and 161.03 deaths every 100,000 inhabitants (Coronavirus Resource Center, 2021).

Despite the extraordinary speed at which vaccines have been developed, Latin America has been slow in rolling them out. The estimation is that a sufficient level of immunity can only be reached around the first half of 2023 (BBC, 2021a, b). As a result, governments are cautiously prepared to release and tighten lockdown interventions throughout 2021 and probably 2022.

In view of the public health hazard, traditional classes have been affected at all levels of education. Many governments decided to close schools nad have kept them closed for year, consequently having to shift suddenly to online and distance learning.

In addition to school closures, learners' households and the education system have also been affected by the economic downturn and restrictions on mobility to enable social distancing. The economic impact of the pandemic will be significant because the largest economies in the region were already growing sluggishly before the outbreak, and the lockdown policies have aggravated the situation. According to IMF, economic growth was 1.6% in 2019 and the estimation for 2020 is a reduction in the regional GDP of 7.4%. Growth is expected to resume in 2021 and would reach an estimate of 4.1% (IMF, 2021), if vaccination campaigns are sufficiently successful. Figure 9.1 shows the fall in the GDP in Latin American countries.

The impact of the economic downturn is bound to affect the marginalised and vulnerable population groups the most, and some authors claim that poverty levels have returned to those observed a decade ago (Alkire et al., 2021). With poverty and inequality on the rise (Sanchez & García, 2021), ECLAC estimates that all income groups will be affected across the region but some more than others (Filgueira et al., 2020), as shown in Fig. 9.2. The method compares the percentage of the population in 2019 and 2020 with incomes in specific brackets, that could define them as extreme poor, poor, low income above poverty line, low medium income, and

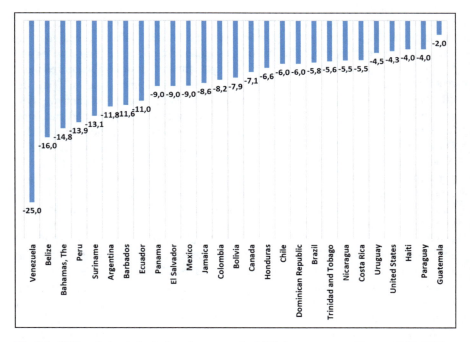

Fig. 9.1 GDP variation in Latin American countries 2020, in percentages. (Source: IMF, 2020)

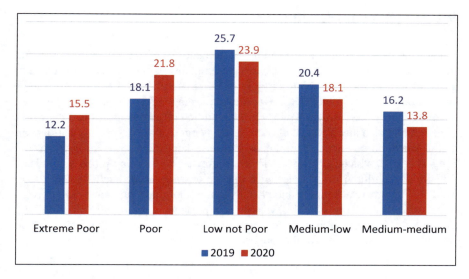

Fig. 9.2 Percentage of the population by income brackets in Latin America. (Source: Filgueira et al., 2020, p. 28)

medium income. The graph shows an increase in the percentage of the population with incomes that fall in the categories of poor and extreme poor, while the percentages of the population classified as not poor and above has decreased.

While the socioeconomic conditions of households generally affect the quality of the learning processes among children and youth, they are the critical factor that causes schooling delays, failure, and desertion. These consequences peak during periods of economic crisis, when poor households withdraw their children from school; the effect is higher for males because they tend to find employment as low skilled labour faster than females (Casquero & Navarro, 2010). Increased schooling failure are bound to occur in Latin America, following the economic slowdown caused by the pandemic and its consequent increase in poverty levels, shown in Figs. 9.1 and 9.2.

In addition, the economic crisis has led governments to apply budget cuts across all national expenses, and education is one of the areas in which they the budget cuts have been the most significant. ECLAC estimates that cuts in the public education budget will be of 9% in 2020 (AFP, 2020). With preliminary data for 2020, the Education Finance Watch 2021 shows that two thirds of the low-income countries in Latin America have reduced their education budgets since the beginning of the pandemic (World Bank and UNESCO, 2021). It is expected that these budget cuts will be in place in 2021 as well.

9.4 Impact on Schooling

One of the most common interventions to mitigate the effect of Covid-19 was the closing of schools and universities. Most governments in the world had to decide to suspend in-classroom teaching partially or completely. A UNESCO study shows that schools have been closed for a worldwide average of 22 weeks up to the week of 25th January 2021. As shown in Fig. 9.3, children of school age around the world have not attended any physical classes for an average of approximately half a year. Latin American countries are the ones that have kept their schools closed for the longest period, which is 29 weeks. This has affected the learning process of at least 137 million children of school age in the region (Seusan & Sachs-Israel, 2020). UNICEF reports in a complementary study that schools have remained closed for an average of 95 learning school days, which represents approximately half an academic year. The research also confirms that Latin America is the region where schools have been closed for the longest time, with an average of 158 days.

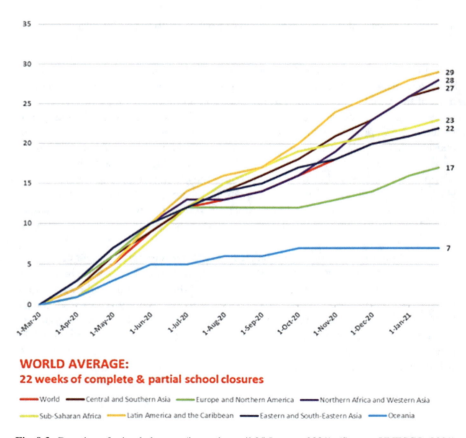

Fig. 9.3 Duration of school closures (in weeks until 25 January 2021). (Source: UNESCO, 2021)

Looking at school closures worldwide, half of the twenty countries that closed their schools the longest are in Latin America. Until March second, 2021, the top of the list includes Panama (211 days), El Salvador (205 days) and Bolivia (192 days). In addition, until March second, 2021 there were 16 countries that closed schools for 100 days or more, and in the group of 20 countries that lost less than 100 days of classes, schools were partially closed for an average of 63 days (UNICEF, 2021). Although the data reflect a severe disruption of the education sector, the crisis is still unfolding and new outbreaks of the pandemic may increase the number of days with school closures.

9.5 Technological Preparedness

While lockdown measures included school closures, governments wanted to avoid the complete cancellation of classes. The efforts hence turned to implementing distance learning by means of virtual environments and new technologies. They were sometimes supported by international organizations such as UNICEF and UNESCO, but there were no contingency plans ready to implement distance learning in any Latin American country before COVID-19 appeared. Distance education by means of digital technologies was extraordinary even at university level and often restricted to expensive programmes for high-income groups.

Although the concepts of distance learning and virtual classes via Internet are currently overlapping, they do not address the same pedagogical modality. Distance learning pre-exists the introduction of Internet and digital technologies and was based on the use of paper, radios, and self-learning. However, the advance of digital technologies has made these methods obsolete and any learning that does not take place inside a classroom in a learning centre is understood to be based on Internet, virtual learning environments and digital platforms.

The turn towards digital learning appeared promising to protect public health in the emergency and, according to some, it was the only safe solution to diminish social contact and the spread of the pandemic. However, the implementation of distance learning did not work as envisioned. Distance learning by means of digital technologies requires a supporting infrastructure grouped under the name of Information and Education Management Systems (IEMS). In the less developed countries in Latin America, a level of leaning based on new technologies could not be achieved, not even close to those gained in physical classes. Most schools did not have digital platforms or any of the other technologies necessary to offer distance learning, so hardly any schools in Latin America have managed to cover the minimum course contents planned for the 2020 school year. Even in the most advanced private schools, where the children of better-off families get their schooling, the infrastructure was not ready to support distance learning.

Information and Education Management Systems include several components: learning platforms, digital content, physical material or social networks, TV or radio, and open schools. A research by the Inter-American Development

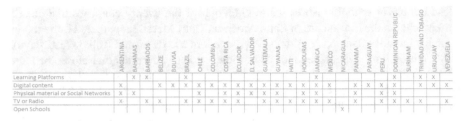

Fig. 9.4 Measures for education continuation. (Source: BID, 2020, p. 6)

Bank (IDB) [Banco Interamericano de Desarrollo (BID)] studied the level of preparedness for distance learning of Latin American public schools in 26 countries. Figure 9.4 shows the various components necessary to implement distance learning and the limitations faced by the schools' systems in Latin America.

The IDB [BID] report studied the degrees of preparedness in these various elements and coded them with the colours green to red. It concluded that only Uruguay was in a situation to implement distance learning based on digital technologies. Other countries, such as Peru and Bahamas, were already advanced with some of the components of an information and education management system, and none of elements appear in red while some factors appear in green. All the other countries included in the report show at least one completely missing component, in red. In the cases of Bolivia, Haiti, Nicaragua, and Venezuela, the IEMS were not developed at all and all appear in red, showing that it was impossible to implement virtual classes in these countries. The colour-coded list is shown in Fig. 9.5.

At the same time as there are important differences in the levels of technological preparedness to implement distance learning, there are also significant inequalities within the countries. Public schools, especially those located in rural and poor areas, lacked the most basic material conditions, such as equipment, computers and Internet connections, to shift to digital learning. In cases where reasonable material conditions are present, these may not be homogeneous and stable enough to facilitate digital learning. Finally, an additional barrier to digital learning is the teaching staff who often lacks the necessary training to teach their learnings in a digital environment.

The inequality in the access to learning via digital platforms between high- and low-income groups is captured by the concept of digital divide or digital gap (Wei & Hindman, 2011; Zheng & Walsham, 2021). The concept initially addressed the inequality in material access to digital technologies but as Internet became increasingly widespread, there was a conceptual shift to actual use and the social consequences of this gap. Wei and Hindman (2011) hence argue that the digital divide can be better defined as inequalities in the meaningful use of information and communication technologies.

Schooling in the Covid-19 pandemic depends critically on both technological access and meaningful use, so the digital gap varies in this context between countries and within countries. Schools may have the necessary technological infrastructure to implement digital learning, but the digital gap may still occur at the household

9 Effects of COVID-19 on Education and Schools' Reopening in Latin America 129

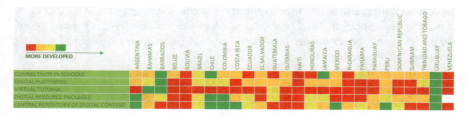

Fig. 9.5 Digital situation of IEMS. (Source: BID, 2020, p. 10)

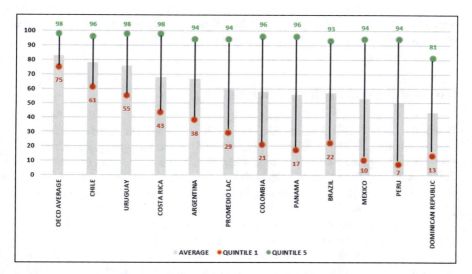

Fig. 9.6 Computer access at home for school homework by socioeconomic level PISA 2018. (Source: BID, 2020, p. 14)

of the students. For instance, a household may have only one computer or laptop but several children or lack sufficient connectivity to receive classes online, especially when there are several children trying to take classes digitally. In turn, technological infrastructure does not imply capacity to use teaching materials and effectively learn from them.

A study by the BID (2020) shows the inequalities in access, not in usage, among households for the top and bottom quintiles, as shown in Fig. 9.6. It shows that the top quintile has very similar levels of access across countries, except in the Dominican Republic. About 95% of the better-off households have computers and connectivity at home, as a condition for digital learning. In contrast, there are marked differences among the marginalised learners of the different countries. A child in a poor household in Chile seems to have better chances of following virtual classes than a child of a similar socioeconomic background in Peru, for example, where the basic infrastructure for home schooling is present in only 7% of the households. Clearly, these learners will miss the opportunity of continuing their education during school closures by means of digital learning (UNESCO, 2021). As a point of reference, in OECD countries 75% of the children have access to a computer and some governments provided free laptops to learners in disadvantaged households (Whittaker, 2020).

In other words, the digital gap has its roots in social inequality and it is reproduced simultaneously with inequalities at large. The digital gap in the less developed countries in Latin America will expand the differences between rich and poor. While there is some evidence that digital schooling can be as effective as physical classes (Garcia & Ruiz, 2010), no distance learning takes place without access to the technologies that make that possible. In addition, the actual learning outcomes depend on use and not only access, and there is minimal information of effective use of digital technologies in home schooling in Latin America.

9.6 Additional Social Consequences

In addition to missing schooling time and not learning specific contents, months of school closures are showing other effects. Children are losing study methods and routines that they had already gained when they could attend classes. Social distance is affecting their learning capacities, which will hinder their employability in the future. Considering the economic impact of the pandemic and the increase in poverty, the prospects of any reduction in inequalities seem sombre.

There are other consequences of school closures that go beyond what children do not learn in terms of contents. Schools often serve as a source of health, nutrition, and social support for those from lower socio-economic segments. While these additional benefits have attracted children to schools in the past, school closures increase the chances that poorer households may decide to drop out schooling in favour of adding workers to earn a livelihood and put food on the table. In terms of health, social contact at school had positive effects on children's health, so it can be expected that their closures may have the opposite effect. The consequences on health are still poorly understood. For instance, there is some evidence among children and youth that they are suffering various mental illnesses caused by being at home for extended periods of time, the lack of socialization, long hours in front of screens of computers and cell phones and (Leicester, 2021) reports a significant increase in infant and youth depression.

Additionally, in some locations the schools are the heart of the community, and this is especially the case in marginalized urban places and remote rural areas where other public services are not available or reach less. In a recent project with a group of schools that promote critical education in Latin America in the poorest areas of each country, we found that among vulnerable populations the schools are the meeting points where children feel safe, households build solidarity around the children and information on well-being and needs flows among them.[2] We concluded that in locations where gang violence and criminality are high, schools are often safe havens where feudal conflicts are left out, and this phenomenon was observed in areas in Bolivia or Colombia, where illegal crops are produced and manufactured with the consequent criminality and violence. Children, teachers, and families in the communities referred to schools using terms such as "the heart of the community", "a sanctuary" and "a refuge in the neighbourhood". Even if school closures are temporary, these social benefits are missing in the lives of children and their families during the pandemic.

While lockdown policies were necessary and allowed countries to curtail the rate of Covid-19 infections, our findings cast light on the negative consequences of school closures on social cohesion at the local level and the extent to which it disrupts everyday life of communities. School closures also undermine the chances of the poor making a living, in addition to undermining the community fabric and the inclusion of children and youth in local governance.

9.7 Reopening of Schools?

Considering the effects of school closures in the pandemic, the reopening of schools appears as a priority. UNESCO General Director, Audrey Azoulay, requested to keep "school closures as the last resource and their reopening a top priority" (UNESCO, 2021). However, Latin American countries that rushed to reopen schools without enough preparation soon faced soaring levels of contagions (Seusan & Sachs-Israel, 2020). It was the case of Uruguay, for example, where the government decided to return to physical schooling in July 2020 (Meritxell, 2020) and in February 2021 the number of infections per million inhabitants was higher than in Brazil, one of the epicentres of the pandemic (Lissardy, 2021). Most children and youth are asymptomatic carriers of the coronavirus, which makes them perfect vehicles for its dissemination and as origins of clusters of infections (Uzín, 2020) when they bring the disease to the high-risk members of their households.

The virus has proven to be resistant and dangerous, able to hide for weeks and then reappear with a symptomatic patient, so it has become clear that full schools' reopening will not be possible until the pandemic is contained. The current situation suggests that it schools reopening will not be possible in any Latin American country in the first half of 2021, at least not permanently. Re-openings in the second half of the year would take place selectively, partially and for short-periods, between one wave of Covid-19 infections and the following. This will be the case, at least, until vaccination campaigns reach a threshold that make it possible to allow social contact again. With these prospects, the choices are limited to strengthening current

---

[2] The name of the schools' network cannot be disclosed. The study was conducted between 2018 and 2020 by a team led by the authors of this chapter and it included 4 case studies of educational communities in six countries in Latin America.

efforts to conduct distance learning, most probably based on digital technologies. Save the Children refers to these efforts as the need to "keep learning alive" as long as the pandemic forces governments to keep schools closed (Warren & Wagner, 2020). Other voices have joined in this conclusion, including UNICEF, several governments, civil society organisations, and businesses, as reported in the UNICEF report "Educación en Pausa" (Seusan & Sachs-Israel, 2020).

In general, the conditions to face this pandemic in Latin America are far from ideal. The introduction of digital technologies in education was already part of the pending agenda to modernize schooling, but in most countries this point was never implemented because fiscal resources started falling about the same time as the goal was set. Most Latin American countries are too far in 2021 from the required operative levels that would allow for an effective distance learning schooling. There is little room for optimism that digital learning will be nearly as effective as traditional schooling.

At the same time, some voices have underlined that the pandemic could be conceived as an opportunity to tackle once and for all the modernization of schooling by introducing new technologies, but the timing is not promising for these prospects. Considerable investments in school infrastructure are necessary and these will be constrained by the soaring expenses in public health care and other pressures imposed by the control of the pandemic. In addition, governments, especially those facing elections this year, will be tempted to run additional fiscal packages to stimulate their ailing economies and help those groups falling back into poverty, instead of spending more on schooling. The reduction of the education budget is estimated to have reached 9% in Latin America in 2020 (UNICEF, 2021) and follows a gradual reduction over the previous 5 years, as we explained in Sect. 9.2 of this chapter.

The reopening of schools is necessary but it appears as a major challenge. Most Latin American countries will need to reopen schools before vaccination campaigns guarantee an acceptable level of health safety, which will most likely translate in additional partial and temporary school closures. Various protocols and health safety procedures will need to be planned and formalized to provide clear guidance on how to minimize biological and sanitary risks of children and teachers returning to schools (Bos et al., 2020). These school reopening could occur only when the contagion levels are relatively low, so schools would not become infection centres for their communities. Compliance with these protocols will need to be enforced and monitored, which is surely going to prove a challenge for many governments that have a poor track record in following up the compliance of their policies (Cluster Global Education, 2020).

In a nutshell, we envision that the future will not bring one school reopening but several periods of partial and targeted school re-openings, by region and depending on the specific conditions. In the medium run, the consequences of the pandemic on schooling will grow, as our own thinking about them develops. Governments and society in general are learning to look beyond the specific role of the school as the location where teaching takes place although the roles as community centres and socialisation spaces are still vacant.

In the long run, several international organizations and governments already envision that the consequences of school closures on this generation will be severe. We can only expect a mitigation of these consequences, not a full recovery, which would require a combined effort of various actors at the global, national and local level. In the short run, the question is how learners and teachers can reconnect with each other and schools, while lockdowns and school remain closed due to the COVID-19 pandemic. In the medium run, relaunching national education systems would need households' and communities' participation and commitment to take children away from the streets, metaphorically, and back to schools. This will be particularly difficult during an economic crisis, when poor households desperate need to make a living. The efforts of schools are bound to be enormous, as several international organizations already claim, because the pandemic has affected learners at all levels: psychological, cognitive, social, content knowledge and education quality. To adjust the education systems to the "new normal" requires changes on infrastructure, teaching methods, curricular content and others, all of which would happen under conditions of stringent resources and when other priorities have pushed education out of the top of the list.

References

AFP. (2020). *El gasto en educación en América Latina caerá 9% en 2020, advierten Unesco y Cepal*. Obtained from France 24: https://www.france24.com/es/20200824-el-gasto-en-educaci%C3%B3n-en-am%C3%A9rica-latina-caer%C3%A1-9-en-2020-advierten-unesco-y-cepal. Accessed on 25 Aug 2020.

Alexander, K. L., Alexander, K., Entwisle, D., & Olson, L. (2007). Lasting consequences of the summer learning gap. *American Sociological Review, 72*(2), 167–180.

Alkire, S., Nogales, R., Quinn, N. N., & Suppa, N. (2021). *Global multidimensional poverty and COVID-19: A decade of progress at risk?* (Ophi research in Progress series, paper 61a). Oxford Poverty and Human Development Initiative, University of Oxford. Available at https://ophi.org.uk/rp-61a/. Accessed on 10 Mar 2021

Arsel, M., & Pellegrini, L. (2021). Indigenous people, extractive imperative and Covid-19 in the Amazon. In E. Papyrakis (Ed.), *Covid-19 and international development*. Springer.

BBC. (2021a). *Vacunas contra el coronavirus: ¿cuál es la situación en tu país?* Obtained from: https://www.bbc.com/mundo/noticias-56025727. Accessed on 12 February 2021.

BBC. (2021b). *BBC Mundo*. Coronavirus | Investigación de la BBC: el continente que no cuenta a sus muertos por covid-19 (y el enorme costo de no hacerlo). Obtained from https://www.bbc.com/mundo/noticias-internacional-56182633. Accessed on 21 Feb 2021.

BID. (2020). *La educación en tiempos del coronavirus*. Banco Interamericano de Desarrollo.

Bos, M. S., Minoja, L., & Dalaison, W. (2020). *Estrategias de Reapertura de Escuelas Durante COVID-19*. BID.

Casquero Tomás, A., & Navarro Gómez, M. (2010). Determinantes del abandono escolar temprano en España: un análisis por género. In *Revista de Educación, número extraordinario* (pp. 191–223).

Cluster Global Education. (2020). *Regreso seguro a la Escuela: Una guía para la práctica*. Cluster Global Education.

Cooper, H. E. (1996). The effects of summer vacation on achievement test scores: A narrative and meta-analytic review. *Review of Educational Research, 66*(3), 227–268.

Coronavirus Resource Center. (2021). *Obtained from Johns Hopkins University & Medicine*: https://coronavirus.jhu.edu/data/mortality. Accessed on 14 Mar 2021.

ECLAC. (2020). *Informe sobre el impacto económico en América Latina y el Caribe de la enfermedad por coronavirus (COVID-19)*. Comisión Económica para América Latina y el Caribe (CEPAL). Obtained from https://repositorio.cepal.org/bitstream/handle/11362/45602/1/S2000. Accessed on 10 Mar 2021.

Filgueira F. G., Galindo L. M., Giambruno, C., & Blofield, M. (2020). *América Latina ante la crisis del COVID-19 Vulnerabilidad socioeconómica y respuesta social*. Obined from Comisión Económica para América Latina y el Caribe (CEPAL) Santiado, Chile: serie Políticas Sociales, N° 238 (LC/TS.2020/149).

Garcia, L., & Ruiz, M. (2010). La eficacia en la educacion a distancia: ¿un problema resuleto? *Ediciones Universidad de Salamanca, 22*(1), 141–162.

IMF. (2020). *Macroeconomic and financial data*. International Monetary Fund. Available at: https://data.imf.org/?sk=3E40CD07-7BD1-404F-BFCE-24018D2D85D2. Accessed on 22 Oct 2020

IMF. (2021). *Informes de perspectivas de la economía mundial enero de 2021*. International Monetary Fund. Available at: https://www.imf.org/es/Publications/WEO/Issues/2021/01/26/2021-world-economic-outlook-update. Accessed on 21 Jan 2021

Leicester, J. (2021). *AP NEWS*. Obtained from https://apnews.com/article/global-rise-childhood-mental-health-pandemic-8392ceff77ac8e1e0f90a32214e7def1. Accessed on 12 Mar 2021.

Lissardy, G. (2021). *Cómo Uruguay pasó de ser una excepción en la pandemia de coronavirus al país con mayor tasa de casos nuevos en América Latina*. Obtained from BBC: https://www.bbc.com/mundo/noticias-america-latina-56412203. Accessed on 16 Mar 2021.

Lorente Rodríguez, M. (2019). Problemas y limitaciones de la educación en América Latina. Un estudio comparado. *Foro de Educación, 17*(27), 229–251.

Mariscal, J., Aneja, U., & Sorgner, A. (2019). Bridging the gender digital gap. *Economics, 13*(1), 1–12.

Meritxell, M. (2020). La receta de Uruguay para el reabrir las escuelas. *El Diario de la Educación*. Obtained from https://eldiariodelaeducacion.com/2020/08/05/la-receta-de-uruguay-para-el-reabrir-las-escuelas. Accessed on 5 Aug 2020.

Sanchez, Á., & García, J. M. (2021). *La pandemia acrecienta la desigualdad y la pobreza en América Latina*. Obtained from The Conversation: https://theconversation.com/la-pandemia-acrecienta-la-desigualdad-y-la-pobreza-en-america-latina-155668. Accessed on 28 Feb 2021.

Seusan, L. A., & Sachs-Israel, M. (2020). *Educación en Pausa*. Panamá: Fondo de las Naciones Unidas para la Infancia (UNICEF). Obtained from https://www.unicef.org/lac/media/18251/file/Educacion-en-pausa-web-1107.pdf

Tedesco, J. (2012). *Educación y justicia social en América Latina*. Buenos Aires.

Torres, R. M. (2005). *12 Tesis para el cambio educativo: Justicia educativa y justicia económica*. Fronesis.

UNESCO. (2015). *Situación Educativa de América Latina y el Caribe: Hacia la educación de calidad para todos al 2015*. UNESCO/OREALC/UNESCO Santiago.

UNESCO. (2021). *COVID-19 education response: Preparing the reopening of schools: Resource paper No29*. UNESCO. Obtained from https://en.unesco.org/news/unesco-figures-show-two-thirds-academic-year-lost-average-worldwide-due-covid-19-school. Accessed on 21 Jan 2021

UNICEF. (2021). *COVID-19 and school closures: One year of education disruption*. UNICEF for every child.

Uzín, G. J. (2020). *¡La segunda ola viene vestida de irresponsabilidad!* Obtained from Pàgina Siete: https://www.paginasiete.bo/opinion/2020/12/28/la-segunda-ola-viene-vestida-de-irresponsabilidad-279315.html?fbclid=IwAR3z71tlvDmApyVyzuXUDvl8LuAgvDxkJGqnVsfQUHetzRL7H331zNESgI4. Accessed on 28 Dec 2020.

Warren, H., & Wagner, E. (2020). *Save our education, protect every child's right to learn in the COVID-19 response and recovery*. Save the Children 2020.

Wei, L., & Hindman, D. (2011). Does the digital divide matter more? Comparing the effects of new media and old media use on the education-based knowledge gap. *Mass Communication and Society, 14*(2), 216–235.

Whittaker, F. (2020). *Coronavirus: Free laptop scheme allocations fall short. Schools Week.* Available at: https://schoolsweek.co.uk/coronavirus-85m-free-laptops-scheme-falls-short. Accessed on 1 May 2020.

WHO. (2020a). *World Heald Organization.* Obtained from Novel Coronavirus (2019-nCoV) Situation Report-11, 31 January 2020: https://www.who.int/docs/default-source/coronaviruse/situation-reports/20200131-sitrep-11-ncov.pdf?sfvrsn=de7c0f7_4. Accessed on 10 July 2020.

WHO. (2020b). *World Health Organization.* Obtained from Coronavirus disease (COVID-19) Weekly Epidemiological Update and Weekly Operational Update: https://www.who.int/emergencies/diseases/novel-coronavirus-2019/situation-reports. Accessed on 10 July 2020.

WHO. (2020c). *World Health Organization.* Obtained from Coroavirus disease 2019 (COVID-19), Situation Report-51: https://www.who.int/docs/default-source/coronaviruse/situation-reports/20200311-sitrep-51-covid-19.pdf?sfvrsn=1ba62e57_10. Accessed on 10 July 2020.

World Bank and UNESCO. (2021). *EFW: Education financial watch 2021.* UNESCO. Obtained from https://unesdoc.unesco.org/ark:/48223/pf0000375577. Accessed on 10 March 2021.

Zheng, Y., & Walsham, G. (2021). Inequality of what? An intersectional approach to digital inequality under Covid-19. *Information and Organization, 31*(1), 100341.

Chapter 10
Indigenous People, Extractive Imperative and Covid-19 in the Amazon

Murat Arsel and Lorenzo Pellegrini

Abstract The challenge posed by Covid-19 for indigenous people of the Amazon is formidable. Remoteness and institutional racism compound the obstacles faced by the general population of Latin America. However, it would be wrong to assume indigenous communities are helpless. They have an extensive repertoire of strategies to respond to external threats that can be conceptualized as part of community resilience. At the same time, indigenous resilience has been affected by the presence of extractive industries in their territories. Oil extraction, logging, mining and oil palm plantations are some of the many extractive processes in the Amazon. We discuss indigenous communities' strategies deployed to counteract Covid-19 and the role played by extractive industries. Given the failure of the global community to develop, produce and make rapidly and universally available vaccines, the future of indigenous people with respect to the pandemic is entangled with broader geopolitical dynamics that determine who can access vaccination. In the longer run, the resilience of indigenous peoples to external shocks will be contingent on their autonomy and especially on the control of their territories.

> "We've been fighting for our lives for centuries. Our elders taught us how to fight the rubber tappers and the oil companies and the loggers. Now we need to protect them from this disease. If our elders die now, the youth will lose their way and won't be able to survive against all of the threats" (Anderson 2020).

10.1 Introduction

The Covid-19 pandemic poses a formidable challenge for indigenous peoples of Latin America, in the Amazon and beyond. For Latin America as a region, throughout the pandemic, varying levels of contagion have met health services of uneven

---

M. Arsel · L. Pellegrini (✉)
International Institute of Social Studies (ISS), Erasmus University Rotterdam,
Rotterdam, The Netherlands
e-mail: pellegrini@iss.nl

© The Author(s), under exclusive license to Springer Nature
Switzerland AG 2022
E. Papyrakis (ed.), *COVID-19 and International Development*,
https://doi.org/10.1007/978-3-030-82339-9_10

quality that at times have been overwhelmed.[1] In the region, indigenous people face compounding challenges associated with the lack of availability of health services in remote locations and with entrenched racism (Goha et al., 2021; Orta-Martínez & Finer, 2010). Moreover, indigenous people already suffer from disproportionally high levels of (extreme) poverty, experience a higher burden of disease and are exposed to intergenerational epigenetic stressors that increase morbidity and mortality rates (Curtice & Choo, 2020; Kaplan et al., 2020). At the same time, there are political economy dynamics that, while many economic activities went into lockdown and experienced blockages, allowed the continuing operation of extractive industries (OECD, 2020). These industries have effectively added the spread of Covid-19 to the list of socio-environmental impacts that the extractives produce in the Amazon (Menton et al., 2021; O'Callaghan-Gordo et al., 2016; O'Rourke & Connolly, 2003; Orta-Martínez et al., 2018; Pellegrini & Arsel, 2018). The disease's impacts on indigenous communities are likely to be particularly severe but it would be erroneous to assume that they are helpless victims. They can rely on ancestral practices and have developed extensive knowledges, tools and strategies that can strengthen resilience vis-à-vis the pandemic.

In this chapter, we discuss the existing vulnerabilities of indigenous people to Covid-19 within a broader historical context before highlighting key aspects of their resilience embodied in concrete strategies employed to limit the spread of Covid-19 and deal with its consequences. We then explore the role played by extractive industries in the spread of the pandemic. Finally, we look at the prospects offered by vaccination for indigenous people and reflect on the possibility of their securing meaningful autonomy within their territories.

10.2 Covid-19, Vulnerability and Resilience

The vulnerability of indigenous people to Covid-19 is shaped and compounded by their already existing marginalization. While these conditions have their roots in the colonial encounter centuries ago, the specific shape of coloniality and its associated dynamics of racism and socio-economic marginalization have changed over time (Galeano, 1973). This section discusses these with specific reference to Ecuador's Amazonian indigenous communities, whose experience shows similarities with indigenous communities of the rest of the Amazon Basin.

For much of colonial and independence-era, the Ecuadorian Amazon experienced neglect from the state. This was due partly to geographical isolation from urban centres and political power. More important, however, were the ways in which Ecuadorian elites looked down on indigeneity through thinly disguised racism, seeing them at best as noble savages and at worse as members of a backward and stagnant culture in need of civilizing (Radcliffe & Westwood, 1996). As such, beyond

---

[1] See, Healthcare Access and Quality Index (2021).

being seen and defended as part of Ecuadorian territorial space (in episodic border wars with Peru), the Amazonian provinces and their indigenous inhabitants had until recently received scant attention from the continuing Ecuadorian state-building project (Sawyer, 2004).

To the extent that Amazonian indigenous communities had a sustained relationship with external actors, these had until the 1950s primarily been Christian missionaries. While there exists considerable variation amongst different missionary organizations (with the Summer Linguistic Institute marking the nadir of indigenous people's experience with them), missionaries' view of indigenous culture covered a similar spectrum as that of Ecuadorian elites (Kimerling, 2013). From 1950s onwards, however, the role of missionaries took an added significance as they were instrumental in 'opening up' the Ecuadorian Amazon as newly discovered oil reserved necessitated secure and continuous access, not only by the state but also by multinational corporations (Finer et al., 2009). The in-roads made by missionaries paved the way towards the exploitation of oil and brought with it not only in-migration but also sustained interest from the state itself. This attention was necessary to facilitate and expand the operation of the oil industry carried out primarily by European and North American corporations until recently, when Chinese state-owned ones came to play a major role (Fernandez Jilberto & Hogenboom, 2010). It served also to secure (limited) consent from indigenous communities as well as the *colonos* (literally, colonizers) who moved to the area attracted either by employment prospects in the oil sector or to open up agricultural land by occupying and clearing the ancestral land of indigenous communities (Mena et al., 2006).

The legacy of these processes come together in Ecuadorian Amazon's ongoing centrality in national oil economy, resulting in a set of interlocking dimensions of vulnerability rooted in the severe limitation of indigenous territorial control. On the one hand, indigenous communities of the region lost both usage rights of part of their territories (especially to *colonos*, who settle on and clear indigenous land to pursue agricultural production). On the other, they also lost their control of access as the state – who owns all underground resources – gave permission to oil companies to operate in indigenous territories. The arrival of both *colonos* and oil workers in indigenous territories consequently changed the character of the population in the region, creating urban centres who came to dominate it politically, economically and culturally. In terms of the latter, the increasing necessity of indigenous people to interact with outsiders has had a negative effect on their ability to maintain their cultural practices and transmit them to future generations. To the extent that the oil economy and growing urbanization in the region brought with it state institutions, indigenous people have scarcely benefited. Their access to basic infrastructure (e.g. clean water) and basic social services (especially schools and medical facilities) remain woefully inadequate. Urbanization, expansion of the agrarian frontier and the continuing encroachment of oil extraction only serve to exacerbate these vulnerabilities.

Facing the pandemic, indigenous people can rely on and renew a repertoire of knowledge and practices that have defined their resilience on the face of continued socio-economic and cultural marginalization shaped by colonization and racism

(Athayde et al., 2017). We follow the classic definition of resilience: "the persistence of systems and of their ability to absorb change and disturbance and still maintain the same relationships between populations or state variables" (Holling, 1973, p. 14) and realize that while resilience is uneven across communities, the very existence of indigenous communities is in itself a manifestation of resilience. It is also important to recognize that resilience can often serve as an excuse not to attend to the task of dismantling the underlying political economic dynamics of marginalization that call for resilience in the first place.

Resilience-enhancing strategies vis-à-vis Covid-19 in Northern Ecuadorian Amazon can be understood in three interrelated dimensions: access, health and support. The first describes the ability and willingness of communities to regulate human movement (e.g. workers in extractive sector; poachers, etc.) within indigenous territory. Indigenous people have been actively following strategies widely adopted across various scales globally and known to be effective, namely lockdowns, limits to travel and access to strangers and in general attempts to decrease contact within and outside communities (Kaplan et al., 2020). They have also been promoting food self-sufficiency and decreased reliance on markets. Since they lack formal political sovereignty over their territories, the implementation of such measures have been partial at best. This is because communities can exercise control over their own members' movement and behaviour (with some completely forbidding their members from traveling and working outside their territory) but have little sway over those of outsiders, especially those working in the extractive sector. The second dimension, health, describes indigenous communities' ability to (re) develop and deploy 'traditional medicine' to improve community health. This has mainly meant the adaption and use of traditional medicine to prevent and treat Covid-19 infections. While these attempts are highly original and likely to be promising given the depth of indigenous knowledge of tropical medicinal plants, the epidemiological evidence on their effectiveness is scant at best (Walters et al., 2021). Whereas the first two aspects of resilience leverage indigenous peoples' own resources and capabilities, the third leverages their ability to mobilize 'external' support, which in this case can be found in national, regional and international bodies. More specifically, indigenous people can access (financial) resources, scientific knowledge and (medical) supplies from a variety of national, regional and international bodies such as ministries, NGOs and intergovernmental organizations.

10.3 Extractive Industries, Vulnerability and Resilience

As the above discussion already suggests, the exposure to Covid-19 and the capability of indigenous communities to deploy strategies enhancing resilience are, in part, shaped by exposure to extractive industries, defined broadly to include the extraction of hydrocarbons and minerals as well as agro-ecological commodities such as timber and oil palm. It has been argued that natural resource abundance acts as a curse on socio-economic development outcomes (Alfonzo, 2016; Papyrakis &

Pellegrini, 2019; Uslar Pietri, 1936). While the 'resource curse' is an established, if qualified, phenomenon at the national level, its manifestations at the local level have not been investigated extensively and are likely to be more complex (Arsel et al., 2019). Furthermore, the impact of resource extraction on health outcomes, especially for indigenous communities, remains poorly understood but mounting evidence suggests that the environmental liabilities associated with extractive industries pose substantial health risks for local communities and especially for those, like indigenous people, who rely proportionally more on the local environment for their livelihoods.

The impact of extractive industries by facilitating the spread of communicable diseases is well established and the Covid-19 pandemic is the latest reiteration of a common dynamic. The Yanomami in Brazil are an example of the gold mining induced spread of the virus to remote populations (Fellows et al., 2020; Menton et al., 2021). The dynamic is common in the Amazon since the region coincides with various types of (largescale and artisanal) metallic mining and hydrocarbons, that are often illegally extracted and always poorly regulated.

The local impact of extractive industries can be expected on each of the resilience-enhancing strategies. In terms of access, extractive industries operate mostly with non-indigenous personnel and limit the capability of indigenous people to control movement in their territories. In terms of health: extractive industries are one of the causes of urbanization and the spread of the market economy, which disrupts the maintenance and intergenerational transmission of traditional knowledge. They also often result in deforestation, which in turn affects biodiversity, diminishing access to biota sources of traditional medicine. In terms of support, extractive activities have historically led to conflicts between (sub)national governments and the communities. These conflicts can have a negative impact on indigenous communities and on their capability to organize and access external actors especially if state policies turn repressive. However, mobilization against extractive processes can also increase community cohesion and organizational capacity that can be then deployed for several purposes. Moreover, companies themselves might facilitate and contribute to the operations of some external actors, such as public health services, or provide care themselves as part of corporate social responsibility projects. Thus, extractive companies can leverage the weakness of the state to become the sole provider of basic services.

10.4 Vaccine Capitalism, Nationalism and Geopolitics

The marginalization of indigenous communities is unlikely to be corrected in the short- and mid-term, not only because there exists limited political will across the region but also because doing so would require enacting long-term transformations of economic, political and cultural structures. In the meantime, therefore, a short-term solution that can help defend indigenous communities from a Covid-19 triggered genocide as argued by Carlos Nobre from the Brazilian Academy of

Sciences (Harris & Schipani, 2020) can be found in widespread and timely vaccination.

However, existing global dynamics in vaccine production and distribution are major barriers to the vaccination of indigenous people. On the one hand, a 'vaccine nationalism' has emerged around the world and the Trumpian slogan 'America first' resonates in the race to secure a large share of the limited existing supply especially by rich countries and regional blocks such as the European Union. In this context, and given the current failure of hegemonic blocks to provide for the global good, the competition and hope comes from the polycentric geopolitical moment. In particular, China and Russia too have developed their own vaccines and seem more willing to share doses with the global South.

On the other hand, 'vaccine capitalism' made it possible for pharmaceutical companies to receive substantial public funds for the development of the vaccines, while retaining patents and future profit streams in private hands. The elephant in the room of vaccine capitalism remains the scarcity of vaccines and the unwillingness of countries to use existing WTO Trade Related Intellectual Property Rights (TRIPS) rules that allow for the suspension of patents on the basis of security exceptions (Abbott, 2020), or change the rules to accommodate also the sharing of production technologies (Abbott & Reichman, 2020; Wolfe, 2020).

Of particular importance for Latin America is the potential of the vaccines developed in Cuba and have been nicknamed 'vaccines for the Global South' (Yaffe, 2021). As of the end of March 2021, Cuba is the only country in Latin America where vaccines have been developed; moreover, out of the 23 vaccines that have reached phase III clinical trials globally, two are produced by the fully state-owned Cuban biotech industry. The announced strategy for the Cuban vaccines is to focus on the provision and production of vaccines with and for the global South, a silver lining for Latin American countries that are lagging behind in terms of vaccinations. Coming back to indigenous people in particular, even if and when countries were to secure a sufficient supply of vaccines, the combination of objective difficulties related to remote locations and structural discrimination will most likely represent further challenges to vaccines roll-out.

10.5 The Extractive Imperative

In the longer term, the safety of indigenous communities – both in the face of Covid-19 and other future pandemics – cannot be guaranteed unless they exercise meaningful control over their territories. Without such control, the three aspects of resilience we have discussed above cannot be effectively deployed. The single biggest impediment to the realization of this control is the 'extractive imperative' rooted in the prevailing development model and political economy structures (Arsel et al., 2016; Pellegrini, 2018).

The extractive imperative is the result of the deep-seated belief in not just the necessity but also the unavoidability of continuously deepening extractive processes to achieve national development. While many policy makers across Latin America do acknowledge the historical and even contemporary ill effects of the dependence of national economies on primary commodity exports, they also believe that transcending the uneven development patterns that emerge from a dependence on extractives can only be possible by going further into extractivism rather than abandoning it in the short term (Arsel, 2012). For some, as has been argued in Ecuador, intensified extractivism is a prerequisite to transition towards post-extractivism, where growth can leverage on indigenous knowledge and natural wealth by building a vibrant biotechnology sector that can develop high-value added products such as pharmaceuticals (Menon, 2020). For others, as exemplified by the policies of the recently elected Bolivian President Luis Arce, the solution is not to build a post-extractive economy. Rather their preferred solution is to 'industrialize natural resources' – for e.g. instead of merely exporting lithium, Bolivia aims in the long term to produce electric automobiles – so as to capture a much larger (and fairer) share of the global value chain (Perreault, 2020). Either way, national development strategies in most of Latin America – even when controlled by progressive leaders – embody the teleological primacy of extractive processes. The resulting extractive imperative implies that extractive operations must continue and expand under any circumstances, which currently applies also to the pandemic that has otherwise altered and limited many economic activities.

Suffering from the vice-like grip of extractivism, indigenous peoples whose territories are rich with the resources coveted by national governments have had limited success in opposing the encroachment of extractive processes into their territories. This is not only because their existing constitutional rights are often violated with impunity by national governments and corporations. Even when indigenous movements escalate their demands by engaging in open conflict they are overpowered by national military forces (as was the case in Dayuma, Ecuador (Pellegrini & Arsel, 2018) or Tipnis, Bolivia (Fabricant & Postero, 2015)) or subjected to punitive actions by corporations (as was the case in Andoas, Peru, see Orta-Martínez et al., 2018).

10.6 Conclusions

Looking forward, it is possible to identify a short-term solution to this particular pandemic and a longer-term one that would reduce indigenous people's vulnerability to and resilience against external shocks –including other pandemics that are likely to emerge in the future (Burton & Topol, 2021). The short-term sustainable solution to the pandemic depends on the local and global availability of effective vaccines. The challenge is both national and local, since Latin America has not been able to access the global vaccine market that is highly lopsided in favour of high income countries. Latin America, with the notable exception of Chile, had a very slow start with the roll-out of the vaccine, an outcome of the toxic compounding forces of vaccine nationalism and vaccine capitalism (Kirby, 2021; Prabhala et al.,

2020). Geopolitical competition might ameliorate the combined failure and South-South cooperation might make vaccine available sooner than the global market would allow – in particular vaccines from China, Russia and Cuba. Still, extensive vaccine coverage of indigenous people, if compared to the general population, will face the additional challenges that marked also the public health approach to the Coronavirus spread and indigenous people will most likely be disadvantaged because of remoteness and discrimination.

In the longer term, it is necessary to recognize indigenous peoples' rightful claims on their ancestral territories so as to strengthen their power against external actors, be they nation-states or corporations. Without effective control over their territories, indigenous people will remain vulnerable to external threats such as Covid-19. Continued and intensified penetration of extractive dynamics, the ongoing extractive imperative, also serve to undermine their resilience which include their ability to control access to their territories, adapt and deploy their own medicinal knowledge and mobilize external support where necessary.

References

Abbott, F. M. (2020). *The TRIPS agreement article 73. Security exceptions and the COVID-19 pandemic* (Research paper 116). South Centre.
Abbott, F. M., & Reichman, J. H. (2020). Facilitating access to cross-border supplies of patented pharmaceuticals: The case of the COVID-19 pandemic. *Journal of International Economic Law, 23*(3), 535–561. https://doi.org/10/gjgnmg
Alfonzo, J. P. P. (2016). Hundiéndonos en el excremento del diablo. *Año 4/N° 7/Enero-Junio/2016, 1950*(1960), 1970.
Arsel, M. (2012). Between 'Marx and markets'? The state, the 'left turn' and nature in Ecuador. *Tijdschrift voor Economische en Sociale Geografie, 103*(2), 150–163.
Arsel, M., Hogenboom, B., & Pellegrini, L. (2016). The extractive imperative in Latin America. *The Extractive Industries and Society, 3*, 880–887.
Arsel, M., Pellegrini, L., & Mena, C. (2019). Maria's paradox and the misery of missing development alternatives in the Ecuadorian Amazon. In P. Shaffer, R. Kanbur, & R. Sandbrook (Eds.), *Immiserizing growth: When growth fails the poor* (pp. 203–225). Oxford University Press. https://doi.org/10.1093/oso/9780198832317.001.0001
Athayde, S., Silva-Lugo, J., Schmink, M., & Heckenberger, M. (2017). The same, but different: Indigenous knowledge retention, erosion, and innovation in the Brazilian Amazon. *Human Ecology, 45*(4), 533–544. https://doi.org/10/gbtt9d
Burton, D. R., & Topol, E. J. (2021). Variant-proof vaccines – Invest now for the next pandemic. *Nature, 590*(7846), 386–388.
Curtice, K., & Choo, E. (2020). Indigenous populations: Left behind in the COVID-19 response. *The Lancet, 395*(10239), 1753. https://doi.org/10/gjhc28
Fabricant, N., & Postero, N. (2015). Sacrificing indigenous bodies and lands: The political–economic history of lowland Bolivia in light of the recent TIPNIS debate. *The Journal of Latin American and Caribbean Anthropology, 20*(3), 452–474.
Fellows, M., Paye, V., Alencar, A., Castro, I., Coelho, M. E., & Moutinho, P. (2020). *They are not numbers. They are lives! COVID-19 threatens indigenous peoples in the Brazilian Amazon.* Amazon Environmental Research Institute. Available at: https://ipam.org.br/wp-content/uploads/2020/06/NT-covid-indi%CC%81genas-amazo%CC%82nia.pdf. Accessed on 12 Feb 2021

Fernandez Jilberto, A. E., & Hogenboom, B. (2010). *Latin America facing China: South-south relations beyond the Washington consensus*. Berghahn Books.

Finer, M., Vijay, V., Ponce, F., Jenkins, C. N., & Kahn, T. R. (2009). Ecuador's Yasuni biosphere reserve: A brief modern history and conservation challenges. *Environmental Research Letters, 4*(3), 034005.

Galeano, E. H. (1973). *Open veins of Latin America: Five centuries of the pillage of a continent*. Monthly Review Press.

Goha, A., Mezue, K., Edwards, P., Madu, K., Baugh, D., Tulloch-Reid, E. E., Nunura, F., Doubeni, C. A., & Madu, E. (2021). Indigenous people and the COVID-19 pandemic: The tip of an iceberg of social and economic inequities. *Journal of Epidemiology and Community Health, 75*(2), 207–208.

Harris, B., & Schipani, A. (2020). Coronavirus corruption cases spread across Latin America. 07 July 2020, *Financial Times*. Available at: https://www.ft.com/content/94c87005-7eb1-47c4-9698-5afb2b12ab54. Accessed on 03 Mar 2021.

Healthcare Access and Quality Index. (2021). *Our World in data*. https://ourworldindata.org/grapher/healthcare-access-and-quality-index. Accessed on 05 Mar 2021.

Holling, C. S. (1973). Resilience and stability of ecological systems. *Annual Review of Ecology and Systematics, 4*(1), 1–23. https://doi.org/10/bctp75

Kaplan, H. S., Trumble, B. C., Stieglitz, J., Mamany, R. M., Cayuba, M. G., Moye, L. M., Alami, S., Kraft, T., Gutierrez, R. Q., & Adrian, J. C. (2020). Voluntary collective isolation as a best response to COVID-19 for indigenous populations? A case study and protocol from the Bolivian Amazon. *The Lancet, 395*(10238), 1727–1734. https://doi.org/10/gjg8nj

Kimerling, J. (2013). Oil, contract, and conservation in the Amazon: Indigenous Huaorani, Chevron, and Yasuni. *Colorado Journal Intional Environmental Law & Policy, 24*, 43.

Kirby, J. (2021, March 10). *How Chile built one of the world's most successful vaccination campaigns*. Vox. https://www.vox.com/22309620/chile-covid-19-vaccination-campaign. Accessed on 10 Mar 2021.

Mena, C. F., Bilsborrow, R. E., & McClain, M. E. (2006). Socioeconomic drivers of deforestation in the Northern Ecuadorian Amazon. *Environmental Management, 37*(6), 802–815.

Menon, G. (2020). Interview with René Ramírez Gallegos, secretary of higher education, science and Technology of Ecuador during the period of 2011 to 2017. *Revista Ibero Americana de Estudos em Educação, 15*(4), 2126.

Menton, M., Milanez, F., de Andrade Souza, J. M., & Cruz, F. S. M. (2021). The COVID-19 pandemic intensified resource conflicts and indigenous resistance in Brazil. *World Development, 138*, 105222. https://doi.org/10/gjg8w5

O'Callaghan-Gordo, C., Orta-Martínez, M., & Kogevinas, M. (2016). Health effects of non-occupational exposure to oil extraction. *Environmental Health, 15*, 56.

O'Rourke, D., & Connolly, S. (2003). Just oil? The distribution of environmental and social impacts of oil production and consumption. *Annual Review of Environment and Resources, 28*(1), 587–617. https://doi.org/10/dghs9m

OECD. (2020). *The impact of Coronavirus (COVID-19) and the global oil price shock on the fiscal position of oil-exporting developing countries*. OECD.

Orta-Martínez, M., & Finer, M. (2010). Oil frontiers and indigenous resistance in the Peruvian Amazon. *Ecological Economics, 70*, 207–218.

Orta-Martínez, M., Pellegrini, L., & Arsel, M. (2018). "The squeaky wheel gets the grease"? The conflict imperative and the slow fight against environmental injustice in northern Peruvian Amazon. *Ecology and Society, 23*(3), 7.

Papyrakis, E., & Pellegrini, L. (2019). The resource curse in Latin America. In H. E. Vanden & G. Prevost (Eds.), *Oxford Encyclopedia of Latin American politics*. Oxford University Press. https://doi.org/10.1093/acrefore/9780190228637.013.1522

Pellegrini, L. (2018). Imaginaries of development through extraction: The 'history of Bolivian petroleum' and the present view of the future. *Geoforum, 90*, 130–141. https://doi.org/10.1016/j.geoforum.2018.01.016

Pellegrini, L., & Arsel, M. (2018). Oil and conflict in the Ecuadorian Amazon: An exploration of motives and objectives. *European Review of Latin American and Caribbean Studies, 106*(July-December), 209–218.

Perreault, T. (2020). Bolivia's High Stakes Lithium Gamble: The renewable energy transition must ensure social justice across the supply chain, from solar panels and electric vehicles to the lithium extraction that fuels them. *NACLA Report on the Americas, 52*(2), 165–172.

Prabhala, A., Jayadev, A., & Baker, D. (2020). Opinion: Want vaccines fast? Suspend intellectual property rights. 07 December 2020, *The New York Times*. Available at: https://www.nytimes.com/2020/12/07/opinion/covid-vaccines-patents.html. Accessed on 04 Apr 2021.

Radcliffe, S. A., & Westwood, S. (1996). *Remaking the nation: Place, identity and politics in Latin America*. Routledge.

Sawyer, S. (2004). *Crude chronicles: Indigenous politics, multinational oil, and neoliberalism in Ecuador*. Duke University Press.

Uslar Pietri, A. (1936). Sembrar el petróleo. *Revista de Artes y Humanidades UNICA, 6*(12), 231–233.

Walters, G., Broome, N. P., Cracco, M., Dash, T., Dudley, N., Elías, S., Hymas, O., Mangubhai, S., Mohan, V., & Niederberger, T. (2021). COVID-19, indigenous peoples, local communities and natural resource governance. *PARKS, 27*, 47–62.

Wolfe, R. (2020). Exposing governments swimming naked in the COVID-19 crisis with trade policy transparency (and why WTO reform matters more than ever). In R. Baldwin & S. Evenett (Eds.), *COVID-19 and trade policy: Why turning inward won't work* (pp. 165–178). CEPR Press.

Yaffe, H. (2021, March 31). Cuba's five COVID-19 vaccines: The full story on Soberana 01/02/Plus, Abdala, and Mambisa. *LSE Latin America and Caribbean Blog*, London School of Economics. https://blogs.lse.ac.uk/latamcaribbean/2021/03/31/cubas-five-covid-19-vaccines-the-full-story-on-soberana-01-02-plus-abdala-and-mambisa/

Chapter 11
Covid-19 and Climate Change

Agni Kalfagianni and Elissaios Papyrakis

Abstract The Covid-19 outbreak generated an unprecedented drop in global CO_2 emissions as a result of economic disruption and reduced demand for electricity and transportation. Similar to the 2008–9 global financial crisis, several scholars have put forward ideas for a Green New Deal and coordinated green fiscal stimulus that can simultaneously reduce economic hardship (especially for the most vulnerable segments of our societies) and address climate change concerns. During a global health and economic crisis of such proportions, there is a heightened risk that financing of low-carbon investment can fall victim to future budget cuts and reduced international aid. In addition, carbon-intensive travel, trade and tourism jointly contribute to both global warming and the rapid spread of pathogens across borders - forced behavioural changes (as a result of Covid-19 mobility and social contact restrictions) offer a unique opportunity for reflection of our unsustainable lifestyles.

11.1 Introduction

While no countries have been immune to either the COVID-19 pandemic or the unfolding climate change crisis, it is common knowledge that it is the poorest countries and communities that are most exposed to such impacts (Gerard et al., 2020; Oldekop et al., 2020). Lack of institutional preparedness and budget constraints, inadequate infrastructure and limited access to information and public resources constitute a toxic mix that amplifies vulnerability to these global shocks (van der Hoeven & Vos, 2021). This vulnerability extends beyond impoverished local

A. Kalfagianni (✉)
Copernicus Institute of Sustainable Development, Utrecht University,
Utrecht, The Netherlands
e-mail: a.kalfagianni@uu.nl

E. Papyrakis
International Institute of Social Studies (ISS), Erasmus University Rotterdam,
The Hague, The Netherlands

© The Author(s), under exclusive license to Springer Nature
Switzerland AG 2022
E. Papyrakis (ed.), *COVID-19 and International Development*,
https://doi.org/10.1007/978-3-030-82339-9_11

communities in the developing world, which may have limited access to clean water or medical services in case of sickness (see also Mukhtarov et al., 2021). Even within wealthier economies, the impact is not felt equally across all segments of the population – low income households, marginalised ethnic communities, refugees, the elderly, those suffering from chronic illnesses and morbid obesity, all find themselves at increased risk (Tai et al., 2020; Wang & Tang, 2020). Sociodemographic inequalities in exposure to the Covid-19 pandemic (but also to extreme weather) reproduce, to a large extent, existing inequalities (both within and across countries) in income, assets, political representation and access to public services (Murshed, 2021). Although the Covid-19 pandemic has somewhat deflected interest away from climate change issues, it is important that scholarly and policy debates on both issues feed into each other rather than evolve in isolation.

This chapter aims at highlighting how interwoven the two debates on Covid-19 and climate change are. It first discusses how the pandemic helped curb emissions in 2020 and speculates on a quick rebound as the global economy gradually recovers. It then proceeds to discuss common features of both the climate change crisis and the coronavirus pandemic, especially regarding their mutual driving forces and past causes. Then, the focus shifts on policy responses that can both confront climate change, as well as assist in the post-Covd-19 recovery. Last, there is a critical reflection on how forced behavioural changes (as a result of Covid-19 mobility and social contact restrictions) may offer a unique opportunity for reflection of our unsustainable lifestyles (and possibly induce a more sustained change in our habits and preferences).

11.2 Covid-19 and the Science of Climate change (Current and Future Projections of CO_2 Emissions)

One of the few positive side-effects of the pandemic has been the drastic decline in global carbon emissions as a result of the economic downturn, drop in energy demand, repeated lockdowns in affected economies and restrictions in mobility. Air quality, especially in urban areas, also improved due to Covid-19 restrictions and its economic impacts (e.g. see Ming et al., 2020 for China; Singh & Chauhan, 2020 for India; Stratoulias & Nuthammachot, 2020 for Thailand). While initial predictions pointed to a potentially larger drop in carbon emissions (close to 8% globally, see Dafnomilis et al., 2020), the (subdued) economic recovery of the second half of 2020 led emissions to somewhat bounce back (with an overall dip close to 6%, see Tollefson, 2021).

The analysis by Dafnomilis et al. (2020) provides medium-term estimates of carbon emissions up to 2030. They estimate that a prolonged economic downturn can lead to a 7% drop in carbon emissions by 2030 against the pre-Covid earlier forecasts. However, a number of 'rebound effects' can potentially quickly wipe out any initial reductions in carbon emissions – this can be in the form of increased car use instead of public transport as a precaution against Covid-19 infection, an oil and coal bounce back as a result of lower fossil fuel price and/or increased demand in the aviation sector as travel restrictions gradually ease. In this latter pessimistic scenario, any initial reduction in carbon emissions will be fully offset by 2025. Naturally, the evolution of carbon emissions will also depend on the recovery measures adopted by governments around the world and their commitment to combine economic stimulus with carbon abatement. Smith et al. (2021) present even more pessimistic projections based on alternative IMF World Economic Outlook forecasts and simulations of fossil fuel prices and quantities; especially due to strong expected growth in developing economies, they predict carbon emissions to exceed their pre-Covid levels within a two-year period.

Li and Li (2021) also speculate that the initial drop in carbon emissions is likely to be of a rather temporary nature. They show how earlier economic shocks (e.g. the first oil crisis in 1973, the collapse of the USSR, the 1997 Asian financial crisis and the 2007–8 global financial crisis) all led to temporary carbon reductions with subsequent rebounds shortly after. This was largely due to economic recovery plans prioritising employment and income creation with little emphasis placed on energy efficiency and environmental quality.

11.3 Similarities Between the Covid-19 and the Climate Change Crises

A number of articles point out the similarities between the Covid-19 and climate change crises. Heyd (2020) notes that both appear to be an integral characteristic of the Anthropocene, a new geological epoch whose central attribute is the visible human impact on the entire planet. Although Covid-19 has a local natural origin, both climate change and Covid-19 are anthropogenically mediated, amplified and transmitted across the globe, while both can be proven fatal when reaching vulnerable populations. In this context, commentators note broader similar driving forces and past common causes, such as urbanisation, dense transportation networks, and deforestation which increases contact with species and spread of pathogens (C-Change, Harvard 2021, website).

Others point out more to the microeconomic foundations of both crises. For example, Fuentes et al. (2020) argue that Covid-19 and climate change alike, involve an overprovision of a global public bad. In their explanation, the Covid-19 pandemic has this characteristic because it is not excludable since it is highly contagious, and non-rival given that becoming infected with the virus does not prevent other people from also becoming infected. Similarly, climate change is a global public bad because it does not exclude people or countries from suffering its adverse consequences, such as extreme weather. Moreover, both Covid-19 and climate change are characterised by externalities whose correction comes at very high economic and social costs (see also Klenert et al., 2020). In the former case, high rates of infection can lead health systems to collapse which also prevents sick people from receiving treatment. In the latter case, greenhouse gas (GHG) emissions remain in the atmosphere for very long periods, sometimes thousands of years, so both present and future generations are impacted and have to pay the price. Finally, both Covid-19 and climate change seem to be characterised by information asymmetries which can lead to sub-optimal courses of action.

Further, both Covid-19 and climate change threats tend to accrue to economically weak and marginalised collectivities and individuals (Heyd, 2020; see also the corresponding chapters on migration, inequality and the informal economy in this book). In the case of Covid-19, poverty, race and status (e.g. indigenous communities) seem to play a determining role both in getting infected as well as recovering. Likewise, in the case of climate change, droughts, floods, heat waves, sea level rise and ocean acidification will cause important disruptions to all but primarily affect the poor and vulnerable. Both crises, then, are not only physical but deeply political, tightly related to questions of inequality and global and local power asymmetries.

Despite these similarities, however, some scholars note that Covid-19 and climate change are also different. The key difference lies in that, while the Covd19 pandemic is likely to be temporary, climate change is irreversible (Gemmene & Depoux, 2020). In turn, this will influence how human societies will and need to respond to these crises in the longer term, a point that we turn to next.

11.4 Policy Responses that Confront Both Climate Change and Assist in the Post-Covid-19 Economic Recovery

The Covid-19 outbreak generated an unprecedented drop in global CO_2 emissions as a result of economic disruption and reduced demand for electricity and transportation. Similar to the 2008–9 global financial crisis, several scholars have put forward ideas for a Green New Deal and coordinated green fiscal stimulus that can simultaneously reduce economic hardship, especially for the most vulnerable segments of our societies, and address climate change concerns (Engström et al., 2020).

Accordingly, for some analysts, the current moment in time presents a unique opportunity to transition to a sustainable post-Covid-19 world, which will require coordinated responses by policy-makers, business and civil society alike (Rosenbloom & Markard, 2020). Yet, these scholars also remind us that this opportunity will not magically materialise. Previous experiences, such as the 2008 global financial crisis, teach us that political responses tend to focus more on stabilising the status quo in terms of industries, technologies and practices, rather than seizing the opportunity for sustainable transformations. To realise the latter, attention would need to be placed on promoting new infrastructure and business models through tax credits and similar measures, as well as harness the disruption caused by Covid-19 to accelerate the decline of carbon-intensive industries, technologies, and practices.

Concerns about the current framing of the New Green Deal are also raised. The European New Green Deal, for instance, is presented as 'a new growth strategy' which for some analysts is problematic (Bogojevic, 2020). These analysts point out that while public debt is a major concern in the wake of the pandemic, it is not entirely clear how environmental protection more broadly and climate change more specifically feature in this debate. While both Covid-19 and climate change require fiscal stimuli, any recovery that focuses strictly on fixing the market will be inadequate in the long run. Similar to the concerns raised previously, it would be a lost opportunity if the petrochemical industry would be rescued in the road for economic recovery, as the banking sector was rescued in the 2008 crisis. Similarly, questions are raised about rescuing the aviation and maritime industry, which are big GHG emitters and were hit very hard by the pandemic. Adopting a "growth" rhetoric risks falling into these traps, despite the simultaneous attention to a "just and inclusive" transition by the European Commission.

In this context, scholars note that the character of the recovery packages will either entrench further or displace (even partly) the current fossil fuel intensive economic system (Hepburn et al., 2020). They identify five policies with the potential to have a simultaneous effect on both economic and climate change impact metrics: clean physical infrastructure, building efficiency retrofits, investment in education and training, natural capital investment, and clean R&D. They argue that, especially in low and middle-income countries, rural support spending replaces clean R&D in terms of importance. The authors reiterate that unless we set the global economy on a net-zero emissions pathway we will be further locked into the fossil fuel system from which it will become increasingly difficult to escape and with disastrous consequences.

To achieve this goal, the need for ample financing to combat climate change with a force similar to that of the pandemic is underlined by Herrero and Thornton (2020). These authors note that, as governments were able to raise US$8 trillion to combat the spread and effects of Covid-19, they should also be able to find funds of similar magnitude in the coming decades both in order to mitigate climate change but also to address its adverse effects on the livelihoods of highly susceptible people in low and middle income countries. The authors reiterate that, unless we treat climate change with the same urgency as Covid-19, we risk disruptions of incalculable magnitude.

Yet, Klenert et al. (2020) argue that climate change mitigation is a much harder challenge for economic policy-making because it requires deep and lasting transformations of the global economy and not only temporary measures. While most measures introduced to counter the pandemic can be lifted once this danger is over, this is not the case with climate change. The latter transformations need to occur long before climate change reaches catastrophic and irreversible dimensions. As many of the climate change effects will be faced by future generations or in distant foreign locations, however, present day politics are very slow to react.

The increasing public debt as a result of the Covid-19 crisis may even contribute to reduced interest in green climate change mitigating investment. Commentators note that the proposed €750 billion package by the European Union and the US$2.3

trillion package by the United States Federal Reserve for the economy are geared primarily for increasing aggregate demand for energy, transport and agricultural products with heighted pressure on natural resources and the environment (Helm, 2020). Investing in healthcare and fibre communication infrastructures may have higher economic and political priority in relation to green investments in the post-pandemic world.

Indeed, during a global health and economic crisis of such proportions, there is a heightened risk that financing of low-carbon investment can fall victim to future budget cuts and reduced international aid. Climate change might be seen as less important in relation to economic recovery, which may create a huge setback to current efforts to reduce GHG emissions (Ecker et al., 2020). The decline of public interest in climate change is also highlighted in the drop of media coverage to climate change issues in 2020 in favour of Covid-19 as a study in Finland demonstrates (Lyytimäki et al., 2020). However, the authors of this study also remind us that fluctuations in media coverage of climate change are typical and it is likely that interest will pick up again in the future, especially around big climate conferences and similar high profile events.

Others scholars emphasise that the main political attention should be placed on pre-disaster preparedness instead of disaster response, as is currently the case, because diseases similar to Covid-19 will very likely appear again in the future (Phillips et al., 2020). This means that adequate technical assistance and funding needs to be devoted to warning systems and international cooperation on a scale commensurate with these global risks. At the same time, technical core capacities need to prioritise equitable outcomes and accountability. Given the unequal distribution of risks to poor, vulnerable and marginalised populations by both Covid-19 and climate change challenges, structural inequalities need to be urgently addressed. Universal access to healthcare is proposed here as a key prerequisite for rectifying both preventative and acute care, emerging from pandemics and climate-related disasters.

Similarly, Newell and Dale (2020) argue for integrated approaches to planning and policy, involving holistic perspectives that recognise interrelationships between different critical sustainability issues and how certain strategies can achieve multiple co-benefits. They contest the conventional economic approach of centralised production which rests on efficiency and economies of scale arguments, while decreasing local and national capacities for resilience. Accordingly, they underscore the benefits of strategically optimising local production-consumption and global supply chains. They urge for a recalibration between local and global production, which will have multiple benefits including associated decreases in GHG emissions. For such a shift to occur, however, global coordination and cooperation is deemed necessary.

11.5 Covid-19, Climate Change and Shifts in Personal Behaviour

The Covid-19 pandemic generated an abrupt pause to a long-term trend of increased carbon-intensive (long-distance) travelling, both for work and pleasure. Most of us had to switch to teleworking (as a result of social distancing rules) instead of commuting long distances to our office space. In several occasions, customers were forced to buy locally and limit prior 'indulgent consumerism' based on unnecessary 'fun shopping'. These forced behavioural changes provided a unique opportunity to reflect on our unsustainable lifestyles and the kind of future economy we wish to co-create after the crisis is over (Bodenheimer & Leidenberger, 2020).

Along these lines, there is intense discussion both within academic and corporate circles on how the current pandemic can permanently change the way we work. The current crisis prompted a rapid switch to a digital economy; most employees (especially those in the service sector) transitioned by necessity to remote work and virtual conference platforms in order to comply with social distancing regulations. However, little attention is often devoted to the digital gap between developing and developed economies (and the associated repercussions for employees in low-income nations that have neither the technical means nor the knowledge to adopt teleworking practices). While almost one in two employees can take advantage of digital technologies to work from home in advanced economies (with the percentage being much higher for certain professions, as in the case of IT technicians, administration professionals and legal experts, see Dingel & Neiman, 2020), Gottlieb et al. (2020) estimate that less than 10% of the labour force could successfully work remotely in developing economies (based on data from Brazil, Costa Rica and Peru). So, while a rise in remote working can simultaneously raise labour productivity and reduce carbon emissions (by minimising unproductive use of time, as a result of carbon-intensive commuting), more public resources and aid need to be devoted towards bridging the digital divide both within societies and across countries (Aissaoui, 2021; Lai & Widmar, 2021).

Goffman (2020) reflects on how the pandemic can provide a critical re-evaluation of our unsustainable living patterns with the emergence of a *glocalization* process. By this, he envisages a global economy that remains interconnected, however, with individuals living more locally as a result of increased awareness and sensitivities towards common worldwide environmental challenges. Along these lines, preferences should gradually (but permanently) shift towards the consumption of locally produced goods, reduced flying and greater use of public transport; this shift needs to be internationally synchronised across the world given the necessity for a coordinated global response to climate change and other global environmental challenges. Many individuals are realising (by necessity) how beneficial local walking and cycling trips are to one's mental health and wellbeing, while also rediscovering the intrinsic value of their local environmental amenities and surroundings (Dzhambov et al., 2020; Xie et al., 2020).

One may naturally wonder whether such a sustained shift in personal behaviour is likely to be wishful thinking. Sheth (2020) claims that consumers do not need to revert to bad old habits, once the pandemic is over (in the same way that new technologies in the past allowed consumers to permanently change preferences and enjoy, for instance, sports and entertainment events from the comfort of their home). After all, psychological experiments reveal that it only takes approximately 66 days for new habits to form and become part of our daily routines (Lally et al., 2010). For many consumers, the pandemic has offered them for the first time the opportunity to enjoy certain services at home rather than from a distance (e.g. attending religious services or conference presentations via live streaming) and it is likely that many individuals will permanently embrace this change in habits (see Dein et al., 2020).

The pandemic imposed a rather abrupt change in human behaviour within a very short period of time; whether some of these environmentally-friendly changes persist in the future (or not) will depend on how convinced the public will be about their long-term necessity. Howarth et al. (2020) emphasise the need for a 'social mandate' where individuals, firms and the government jointly support a similar set of behavioural changes and environmental objectives (e.g. with the help of citizen assemblies/conventions that facilitate the convergence of interests across a wide range of stakeholders). In this way, behavioural changes will be based on consent and informed choices (rather than being superimposed and enforced through government interventions (Smith & Hughes, 2020).

11.6 Conclusions

Most of us experienced 2020 as the most disruptive year we can recall. At the same time, the pandemic response has highlighted how international coordinated action is crucial in tackling global crises. Countries had to simultaneously restrict international (and even local) mobility to prevent the further spread of the virus; they also promptly shared epidemiological knowledge on transmission and control with other governments and scientific bodies. Clearly, the international coordination in the combat against the pandemic was, by no means, a flawless and frictionless process; this is evident from the current international grievances arising due to the delayed and unequitable distribution of vaccines.

Viewing climate change as an emerging global crisis will help facilitate the adoption of similar concerted actions internationally (FitzRoy & Papyrakis, 2017). An important message that one can draw from the pandemic (and its handling) is that we cannot tackle problems that transcend borders (as in the case of global warming) without international concerted action. We have multiple success stories on that front, such as the drastic drop in global extreme poverty in recent decades, the protection of the ozone layer and the almost complete elimination of illiteracy in many parts of the developing world. International leadership (that involves not only governments but also civil society groups and the private sector) can help move the climate agenda forward and transform a possible climate disaster into another success story.

References

Aissaoui, N. (2021). The digital divide: A literature review and some directions for future research in light of COVID-19. *Global Knowledge, Memory and Communication*. In Press.

Bodenheimer, M., & Leidenberger, J. (2020). COVID-19 as a window of opportunity for sustainability transitions? Narratives and communication strategies beyond the pandemic. *Sustainability: Science, Practice and Policy, 16*(1), 61–66.

Bogojevic, S. (2020). Covid-19, climate change action and the road to green recovery. *Journal of Environmental Law, 32*, 355–359.

Dafnomilis, I., Den Elzen, M., Van Soest, H., Hans, F., Kuramochi, T., & Höhne, N. (2020). *Exploring the impact of the COVID-19 pandemic on global emission projections*. PBL Netherlands Environmental Assessment Agency. Available from: www.pbl.nl/en/publications/exploring-the-impact-of-covid-19-pandemic-on-global-emission-projections. Accessed on 17 Mar 2020

Dein, S., Loewenthal, K., Lewis, C. A., & Pargament, K. I. (2020). COVID-19, mental health and religion: An agenda for future research. *Mental Health, Religion and Culture, 23*(1), 1–9.

Dingel, J., & Neiman, B. (2020). How many jobs can be done at home? *Journal of Public Economics, 189*, 104235.

Dzhambov, A., Lercher, M., Browning, P., Stoyanov, M. H. E. M., Petrova, D., Novakov, S., & Dimitrova, D. D. (2020). Does greenery experienced indoors and outdoors provide an escape and support mental health during the COVID-19 quarantine? *Environmental Research, 2020*, 110420.

Ecker, U. K. H., Butler, L. H., Cook, J., Hurlstone, M. J., Kurz, T., & Lewandowsky, S. (2020). Using the Covid-19 economic crisis to frame climate change as a secondary issue reduces mitigation support. *Journal of Environmental Psychology, 70*, 101464.

Engström, G., Gars, J., Jaakkola, N., Lindahl, T., Spiro, D., & van Benthem, A. A. (2020). What policies address both the coronavirus crisis and the climate crisis? *Environmental and Resource Economics, 76*, 789–810.

FitzRoy, F. R., & Papyrakis, E. (2017). *An introduction to climate change economics and policy*. Routledge.

Fuentes, R., Galeotti, M., Lanza, A., & Manzano, B. (2020). Covid-19 and climate change: A tale of two global problems. *Sustainability, 12*(8650), 1–14.

Gemmene, F., & Depoux, A. (2020). What our response to the Covid-19 pandemic tells us of our capacity to respond to climate change. *Environmental Research Letters, 15*, 101002.

Gerard, F., Imbert, C., & Orkin, K. (2020). Social protection response to the COVID-19 crisis: Options for developing countries. *Oxford Review of Economic Policy, 36*(Supplement), S281–S296.

Goffman, E. (2020). In the wake of COVID-19, is glocalization our sustainability future? *Sustainability: Science, Practice and Policy, 16*(1), 48–52.

Gottlieb, C., Jan Grobovšek, J., Poschke, M., & Saltiel, F. (2020). Working from home in developing countries. *European Economic Review, 133*, 103679.

Helm, D. (2020). The environmetal impacts of the Coronavirus. *Environmental and Resource Economics, 76*, 21–38.

Hepburn, C., O'Callaghan, B., Stern, N., Stiglitz, J., & Zenghelis, D. (2020). Will Covid-19 fiscal recovery packages accelerate or retard progress on climate change? *Oxford Review of Economic Policy, 36*(S1), 359–381.

Herrero, M., & Thornton, P. (2020). What Covid-19 teaches us about responding to climate change? *The Lancet, 4*, e174.

Heyd, T. (2020). Covid-19 and climate change in the times of the Anthropocene. *The Anthropocene Review*. https://doi.org/10.1177/2053019620961799

Howarth, C., Bryant, P., Corner, A., Fankhauser, S., Gouldson, A., Whitmarsh, L., & Willis, R. (2020). Building a social mandate for climate action: Lessons from COVID-19. *Environmental and Resource Economics, 76*, 1107–1115.

Klenert, D., Funke, F., Mattauch, L., & O'Callaghan, B. (2020). Five lessons from Covid-19 for advancing climate change mitigation. *Environmental and Resource Economics, 76*, 751–778.

Lai, J., & Widmar, N. (2021). Revisiting the digital divide in the Covid-19 era. *Applied Economic Perspectives and Policy, 43*(1), 458–464.

Lally, P., van Jaarsveld, C., Potts, H., & Wardle, J. (2010). How habits are formed: Modelling habit formation in the real world. *European Journal of Social Psychology, 40*(6), 998–1009.

Li, R., & Li, S. (2021). Carbon emission post-coronavirus: Continual decline or rebound? *Structural Change and Economic Dynamics, 57*, 57–67.

Lyytimäki, J., Kangas, H.-L., Mervaala, E., & Vikström, S. (2020). Muted by a crisis? Covid-19 and the long-term evolution of climate change newspaper coverage. *Sustainability, 12*, 8575.

Ming, W., Zhou, Z., Ai, H., Bi, H., & Zhong, Y. (2020). COVID-19 and air quality: Evidence from China. *Emerging Markets Finance and Trade, 56*(10), 2422–2442.

Mukhtarov, M., Papyrakis, E., & Rieger, M. (2021). Covid-19 and water. In E. Papyrakis (Ed.), *Covid-19 and international development*. Springer.

Murshed, S. M. (2021). Consequences of the Covid-19 pandemic for economic inequality. In E. Papyrakis (Ed.), *Covid-19 and international development*. Springer.

Newell, R., & Dale, A. (2020). Covid-19 and climate change: And integrated perspective. *Cities & Health.* https://doi.org/10.1080/23748834.2020.1778844

Oldekop, J. A., Horner, R., Hulme, D., Adhikari, R., Agarwal, B., et al. (2020). COVID-19 and the case for global development. *World Development, 134*, 105044.

Phillips, C. A., Caldas, A., Cleetus, R., Dahl, K. A., Declet-Barreto, J., Licker, R., Delta Merner, L., Pablo Ortiz-Partida, P., Phelan, A. L., Spanger-Siegfried, E., Talati, S., Trisos, C. H., & Carlson, C. J. (2020). Compound climate risks in the Covid-19 pandemic. *Nature Climate Change, 10*, 586–598.

Rosenbloom, D., & Markard, J. (2020). A Covid-19 recovery for climate. *Science, 368*(6490), 447.

Sheth, J. (2020). Impact of Covid-19 on consumer behavior: Will the old habits return or die? *Journal of Business Research, 117*, 280–283.

Singh, R. P., & Chauhan, A. (2020). Impact of lockdown on air quality in India during COVID-19 pandemic. *Air Quality, Atmosphere and Health, 13*, 921–928.

Smith, G., & Hughes, T. (2020). *Why participation and deliberation are vital to the COVID-19 response*. Available at: www.involve.org.uk/resources/blog/opinion/why-participation-and-deliberation-are-vital-covid-19-response. Accessed on 25 Mar 2021.

Smith, L. V., Tarui, N., & Yamagata, T. (2021). Assessing the impact of COVID-19 on global fossil fuel consumption and CO_2 emissions. *Energy Economics, 97*, 105170.

Stratoulias, D., & Nuthammachot, N. (2020). Air quality development during the COVID-19 pandemic over a medium-sized urban area in Thailand. *Science of the Total Environment, 746*, 141320.

Tai, D. B. G., Shah, A., Doubeni, C. A., Sia, I. G., & Wieland, M. L. (2020). The disproportionate impact of COVID-19 on racial and ethnic minorities in the United States. *Clinical Infectious Diseases, 72*(4), 703–706.

Tollefson, J. (2021). COVID curbed carbon emissions in 2020 – But not by much. *Nature, 589*(7842), 343.

van der Hoeven, R., & Vos, R. (2021). Reforming the international financial and fiscal system for better COVID-19 and post-pandemic crisis responsiveness. In E. Papyrakis (Ed.), *Covid-19 and international development*. Springer.

Wang, Z., & Tang, K. (2020). Combating COVID-19: Health equity matters. *Nature Medicine, 26*, 458.

Xie, J., Luo, S., Furuya, K., & Sun, D. (2020). Urban parks as green buffers during the COVID-19 pandemic. *Sustainability, 12*, 6751.

Chapter 12
Covid-19 and Water

Farhad Mukhtarov, Elissaios Papyrakis, and Matthias Rieger

Abstract The Covid-19 outbreak exerts additional pressure on the global water sector, which is already under much strain as a result of climate change, fast expanding populations, aging and inadequate infrastructure and ill-planned urbanisation (especially in many parts of the developing world). Prolonged water scarcity has also enhanced food import dependencies in arid places and, hence, heightened food insecurity in periods of trade disruption (as in the current Covid-19 worldwide recession). The pandemic will likely have a lasting influence on behaviours, policies and research in the Water, Sanitation and Hygiene sector and beyond.

12.1 Introduction

The Covid-19 pandemic has exacerbated many challenges that were already present and prominent especially in many parts of the developing world (although not exclusively). Limited access to piped water supply has been a chronic concern in many arid environments (and fast-expanding urban centres), which hinders hand hygiene and protection against infectious diseases. While climatic conditions and population dynamics contribute to such water scarcities, poor governance has been recognised as the key driver of failures in the water supply and sanitation sector. Various campaigns and initiatives since the outset of the pandemic have aimed at raising awareness of the importance of access to clean water and sanitation, as well as providing water containers and hygiene kits to vulnerable communities. This has been particularly important for low-income households, crowded slums and refugee camps where the probability of infection and the capacity to treat it remain particularly high. The pandemic and limited access to water can lock vulnerable individuals in a reinforcing vicious circle, where water scarcity leads to higher infection

---

F. Mukhtarov (✉) · E. Papyrakis · M. Rieger
International Institute of Social Studies (ISS), Erasmus University Rotterdam, The Hague, The Netherlands
e-mail: mukhtarov@iss.nl

© The Author(s), under exclusive license to Springer Nature Switzerland AG 2022
E. Papyrakis (ed.), *COVID-19 and International Development*,
https://doi.org/10.1007/978-3-030-82339-9_12

rates, poor health, constrained abilities to earn income and a further reduced affordability of water services.

The chapter aims at highlighting how interwoven Covid-19 and water issues are. It first provides a discussion on broader (short and long term) water security challenges and how these are likely to be amplified by the ongoing pandemic. It then proceeds to discuss how a combination of poor water governance, worsening climatic conditions, increased financialisation of the sector, growing populations and neglect of domestic agriculture have led to rising food-import dependency in many parts of the world. It then reflects on how the pandemic (with its extraordinary disruption to international agricultural supply chains) heightens risks of food shortages. Next, the chapter moves to discuss necessary macroeconomic policies and interventions for the water sector in times of crisis, when it is likely to experience diversion of funds from investment in water-related projects. Naturally, responsibility also lies with each person individually. For this reason, the last section probes into the more micro-developmental and behavioural dimensions in the water, sanitation and hygiene sector.

12.2 Water Security Challenges and the Covid-19 Pandemic

Covid-19 has a serious direct and indirect impact on human societies. Water management practices and their organisation may ameliorate, as well as aggravate such impacts. Major aspects include access to water supply, sanitation, and hygiene (WASH), urbanisation and increased paved surfaces, balanced allocation of water resources among competing users, prevention of water pollution from agricultural run-off and industrialisation, and adequate and sustained funding of in many places dilapidated infrastructure and management (e.g. Butler et al., 2020). In the short-term, the biggest impact of Covid-19 has been on water utilities and disaster relief agencies tasked with the provision of safe drinking water and hygiene to healthcare facilities (WHO and UNICEF, 2020; Kolker et al., 2020). The mutual impact of the pandemic and water is multifaceted and variegated depending on the context. Table 12.1 sketches some of these impacts along the 2x2 matrix based on the nature and the timeline of the impact.

Short-term direct impacts include the inadequacy of the present WASH facilities to prevent the spread of Covid-19 around the world. Hand-washing has been suggested as the primary response to the pandemic (e.g. Neal, 2020). However, "a quarter of all health care facilities have no basic water services, which means 712 million people have no access to water when they use health care facilities" (WHO/UNICEF, 2020: 1). One in three healthcare points worldwide have no facilities for hand-washing (ibid: 1). In lesser developed countries, "half of health care facilities lack basic water services and 60% have no sanitation services" (ibid: 1).

Despite much progress to connect billions to piped water supply in the past two decades, some 25% of the world population still lack adequate water supply (WHO and UNICEF, 2017). Even in the wealthier countries, such as the US, millions live

Table 12.1 Mutual impact of Covid-19 and water management

Impact/Timeline	Short-term impact	Long-term dangers
Direct impact	Water, sanitation and hygiene (WASH) provision to healthcare institutions; WASH provisions in dense (semi)-informal urban areas; Humanitarian supply of (bottled) water to vulnerable populations (e.g. refugee camps); Water utilities' struggle to maintain financial and operational viability	Priority to WASH at the expense of other crucial challenges such as climate change adaptation, land use patterns, ecosystem health and energy; Lesser public funds for water infrastructure and management and the danger of oversized influence of private financial capital in the water sector
Indirect impact	Global supply chains under pressure, including water-intensive crops and products; Competition for water between increasing demands from agricultural and domestic uses; Increasing pollution from agricultural run-off	Securitisation of WASH that may lead to day-to-day and fragmented management; Private financial take-over of water infrastructure and services; Attention to techno-fixes with a relative neglect of social and political aspects of water management and governance

without piped water supply (Tortajada & Biswas, 2020). The urban poor are disproportionately affected and the governments should provide subsidies for bottled water or water trucks in these areas in the short term and piped water supply in the longer term (Neal, 2020; Butler et al., 2020).

While people with no access to sanitation declined from 1229 to 892 million between 2000 and 2015, sub-Saharan Africa has not progressed much (WHO and UNICEF, 2017). In some 20 countries the basic sanitation has actually worsened in this period. These countries require additional support from international donors and charities in order to protect vulnerable populations (see Fig. 12.1).

Covid-19 hit water utilities around the world hard, their revenues have decreased and operational disruptions increased (Butler et al., 2020; Neal, 2020). For example, water companies in Chile and Brazil agreed to postpone or exempt vulnerable households from water payments (Butler et al., 2020). Kolker et al. (2020) suggested to freeze water tariffs and temporarily prohibit disconnections, with financial implications for water utilities. In the U.S., some 57 million people across the country have been allowed to continue receiving water despite their inability to pay for it (Tortajada & Biswas, 2020: 441).

The task of shielding water utilities rests overwhelmingly on the state. However, the private sector can also play a role; Kolker et al. (2020) and Butler et al. (2020) discuss ways of mobilising private capital to finance utilities. There is already much appetite for private sector involvement; Merrill Lynch and the Bank of America estimated that the water industry market could be worth US$800–1000 billion by 2030 (Ahlers & Merme, 2016: p. 768). Furthermore, the World Bank offers various guarantees to investors including assurances against political risks (e.g. Kolker

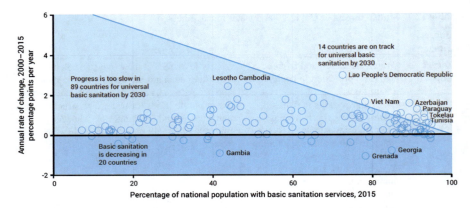

Fig. 12.1 Progress towards universal basic sanitation services (2000–2015) among countries where at least 5 per cent of the population did not have basic services in 2015. (*Source:* WHO and UNICEF (2017, p.14))

et al., 2020). This environment has facilitated an increased competition between industrialised nations and their (private) water sector to occupy the new global water markets (Mukhtarov et al., 2021). However, the pandemic should not become a pretext for a speedy transfer of publicly managed WASH facilities to private hands as it has sometimes been in the past (Klein, 2017). Instead, such transfers must be carefully considered and gradually implemented depending on the regulatory capacity of the governments, and where relevant, include guarantees from donors or the state in case privatisation and financialisation of water backfire (e.g. Bayliss, 2014; Schmidt & Matthews, 2018).

The indirect impacts of the pandemic are mostly felt in the pressures and disruptions of the global flow of goods and services that require much water to produce. As the industrial use of water is expected to drop in the next 2–3 years by 27% and extend across the whole water supply chain, there will be less funds available to support water-intensive industries (Global Water Intelligence cited in Butler et al., 2020). At the same time, the demand for agricultural water use has been steadily growing in the past decades and is likely to continue growing during the pandemic, see Fig. 12.2. The increased demand for water from the agricultural sectors brings two issues to bear on water governance systems, namely:

(a) the serious impact on water quality due to irrigation run-off and soil erosion; and
(b) the lesser availability of water for fast growing cities and changing consumption patterns (e.g. expectations to access safe water 24 h a day).

In terms of indirect and long-term impacts of Covid-19 on water resources, there is a risk of further securitisation of water within the frame of WASH. However crucial for public health and human development, water supply and sanitation services have to be considered within a larger framework of integrated water management,

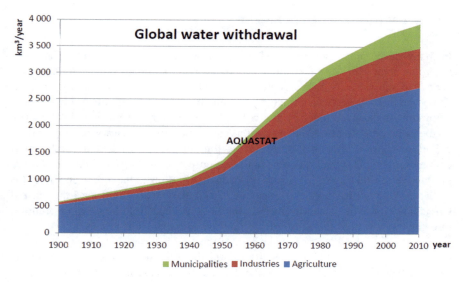

Fig. 12.2 Trends in global water withdrawal by sector between 1990 and 2020 (km^3 per year). (*Source*: FAO (2018))

water quality, ecosystem health and governance (e.g. Gaddis et al., 2019). Securitisation of water is almost always accompanied with an emphasis on the financial gap, and hence calls for more private capital; this may bring about risks in the form of price-hikes, cherry-picking in service provisions, as well as financial speculation in shares (e.g. Bayliss, 2014; Ahlers & Merme, 2016). Whenever private actors participate in the provision of public goods, a strong regulatory framework and capacity of the state is essential (e.g. Mukhtarov, 2007; Akhmouch & Kauffmann, 2013).

Finally, with the growing challenges around water access and safety (partly in view of climate change), technological solutions, such as satellite imagery, water treatment technology and desalination and water reuse technology, have become increasingly important (e.g. AAAS, 2019). The pandemic is likely to provoke the instinct of public managers to cling to technological solutions that seem easy and within reach. However, attention to complex and culturally and politically sensitive issues of water allocation, land use choices, as well as issues around funding infrastructure and planning, remain as important as application of new technologies in the post-pandemic world (e.g. Gaddis et al., 2019; Watkins, 2006). Good water management requires technical as well as adequate governance and managerial approaches, which are also very context dependent. Hence, an experimental and adaptive approach will need to be applied with a clear realisation that there are no universal panaceas to this complex challenge (Ingram, 2013).

12.3 Food Import Dependency

In many developing countries (and often for some of the most vulnerable ones, regarding their ability to feed their fast rising populations), external food dependency has substantially increased during the last few decades. Growing food dependency and food-import bills can exert particular pressure to marginalised low-income communities that face much higher risks of malnutrition and nutrient deficiencies. At the same time, food-dependent nations can avoid food insecurity, as long as their export revenues from other economic sectors suffice to meet their food import demands; even then, this can come at the expense of crowding-out imports of more productive capital goods. In periods of severe disruptions to the global economy (as in the 2007–8 global financial crisis, see Brinkman et al., 2010, and the current Covid-19 worldwide recession, see Laborde et al., 2020), falling export volumes threaten the ability of food-importing countries to remain self-reliant.

Figures 12.1 and 12.2 provide an idea of the significant rise in food dependency over time for some developing nations. Both figures use historical data from the FAO website (FAO, 2021). Figure 12.3 depicts the increase in cereal import dependency between the periods 2000–2 and 2015–7 for Benin, Kenya, Venezuela and Zimbabwe (expressed as the percentage of domestic food supply of cereals that is imported). Figure 12.4 presents a similar pattern for the value of food imports expressed in relation to exports (for the same economies and periods of analysis).

Water scarcity, widely seen in most contexts as the outcome of poor water governance, is likely to be further exacerbated in the near future as a result of climate change and population growth (Watkins, 2006). The cumulative impact of such

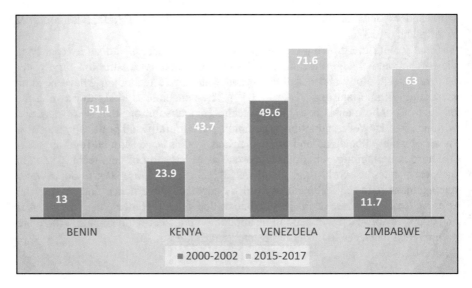

Fig. 12.3 Cereal imports dependency ratio. (Source: FAO (2021))

12 Covid-19 and Water

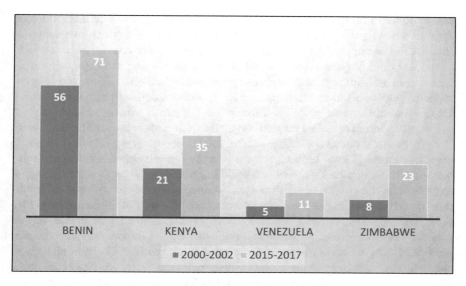

Fig. 12.4 Food imports as a share of export value. (Source: FAO (2021))

scarcity encourages high dependence on food imports for countries located in arid/semi-arid regions (Misra, 2014). Climatic variability and extreme weather incidences also pose additional challenges to the ability of countries to maintain their own food sufficiency (based on local produce). A recent study by Chouchane et al. (2018) predicts that water-scarce economies are expected to become increasingly more reliant on trade to meet their internal demand for staple crops, with a projected increase in trade flows by 40–80% towards 2050 in relation to the 2001–2010 average values.

Naturally, there are many other interlinked factors that have contributed to this higher food-import dependency for certain parts of the world. Over the past decades, trade liberalisation and globalisation resulted in the elimination of many food tariffs and gradual substitution of domestically produced food with cheaper food imports (Otero et al., 2013). This is often perceived as a much more efficient policy in comparison to the much more costly option of subsidising domestic farmers, especially for economies with a positive trade balance that can generally afford such a long-term strategy. Food import dependency is further intensified by patterns of increasing trade specialisation (i.e. economies concentrating efforts and specialising in certain products in exchange for other imported goods, based on their comparative advantage). Specialised production of a limited range of goods can result in efficiency gains and a more productive use of domestic production factors. Even in the case of rural economies, this suggests that specialising in few agricultural commodities can generate increasing returns to scale for the expanding sectors (Costinot & Donaldson, 2012); at the same, specialisation in high-value but water-intensive commodities (as in the case of the floriculture industry in Kenya and Ethiopia, for instance) causes faster depletion of aquifers and constraints the opportunities for conventional farming (Mekkonen et al., 2012). In many cases, reliance on food imports was promoted as a more successful development strategy in relation to small-scale farming, which was largely seen as non-conducive to sustained economic growth. Long-term development plans placed more emphasis on the promotion of manufacturing and technology-intensive sectors, given the overall expectations of declining terms of trade for economies dependent on their primary sectors (i.e. a declining price ratio for the agricultural vs manufactured goods – commonly referred to as the Prebisch-Singer thesis in economics, see Hallam, 2018).

Under normal circumstances, excessive food dependency would not have posed disproportionate risks to food security, especially for vulnerable low-income communities. However, the Covid-19 pandemic has caused an extraordinary disruption to global merchandise trade (with few exceptions, as in the case of medical equipment); early estimates point to approximately a value loss of approximately 20% (UNCTAD, 2020). As the COVID-19 disrupts international agricultural supply chains by restricting transportation and trade flows, food import dependent economies are facing increasing risks of local food shortages (especially in relation to staple food products, as in the case of wheat, rice and maize, see Falkendal et al., 2021). Food shortages are further aggravated by food export restrictions (by producing nations that, as a precaution, stock up reserves for own use), as well as disruption in farming as a result of Covid-19 preventive measures (Benton, 2020). Food inflation, especially for perishable products, in combination with reduced income levels and purchasing ability, impact the poor by limiting access to affordable and nutritious products (Malpass, 2021). These recent developments should raise alarm bells; rising population pressures and accelerating global warming will exacerbate water stress in many parts of the world and intensify external food dependencies. Consequently, this will also further raise the vulnerability of their populations to future disruptions in trade flows that stemm from other impending epi(pan)demics or external shocks.

12.4 Macroeconomic Policies for the Water Sector in Times of Crisis

Even before the Covid-19 outbreak, the global water sector was under much pressure as a result of climate change, fast expanding populations, aging and inadequate infrastructure and ill-planned urbanisation especially in many parts of the developing world (Hanjra & Qureshi, 2010). There are several policy reports predicting a drop in public investment in the global water sector in the years to come (IFC, 2020; World Bank, 2020). As national and local governments (and municipalities) prioritise spending on emergency response and income support for affected communities, there is a high risk that investment in new water projects is likely to be delayed (see also Kalfagianni & Papyrakis, 2021). A recent survey conducted by the Chartered Institution of Water and Environmental Management (CIWEM) with water experts highlighted the common concern of reduced future public investment, cancellation of projects and relocation of centralised funding towards other priorities (Cotterill et al., 2020). Nevertheless, increased interest from private investors, as we discussed in Sect. 12.2, will potentially counterbalance the slowdown in public investment, at least for as long as the economic downturn does not severely constrain the supply of private funds and investment.

The revenues of water utility companies are also likely to be negatively affected, allowing hence less room for future infrastructural investment. In some cases, water consumption increased as a result of increased demand for hand washing; in many cases, however, utilities faced a drop in demand as a result of reduced commuting, lower industrial activity and subdued tourism performance (with revenues falling as much as 50% in some tourism-dependent municipalities; see Cheval et al., 2020).

Given the critical role of handwashing and sanitation in limiting the spread of Covid-19, several countries have resorted to a partial suspension of water billing. These measures were largely adopted as a means to safeguard the uninterrupted use of water for vulnerable communities (and partly mitigate the income loss during prolonged periods of lockdown; see Antwi et al., 2020; Cooper, 2020). Naturally, these are measures in the right direction, aiming at providing some financial relief for those in need, as well as mitigating the spread of the virus. However, they also exert additional pressure to the finances of water utility companies, at least for those cases, where this is unaccompanied by sufficient financial compensation.

Given the combined health and economic crisis of the Covid-19 pandemic, governments (especially those in the developing world) should take a number of alleviating measures. The historically-low interest rates offer an opportunity for countercyclical expansionary fiscal policies that can simultaneously stimulate the economy and lessen chronic shortages of the water sector. Given that public investment in the water sector is often labour intensive, such initiatives would generate employment and reduce the duration and intensity of the economic downturn. The scale of the epidemic and associated human loss has reinforced the need to improve access to safe and affordable water. In addition, while water projects after many decades of neglect are likely to stimulate the economy and prevent the spread of the virus in the short term, they will also help address imminent water shortages as a result of climate change. Clear communication strategies should highlight the multifaceted benefits of such green recovery schemes to the public and, hence, help achieve a broader social acceptance of the urgency in investing in water and other renewable resource projects (OECD, 2020). Governments in developed economies should also resist tempting proposals to cut foreign aid (as compensation for the domestic cost of the pandemic); in an interconnected world, reduced aid to developing countries weakens the fight against the virus both abroad as well as home (Kobayashi et al., 2021). Last, tax reliefs and incentives should incentivise investments in the water sector, especially given the severity of credit constraints many utility companies are likely to face.

12.5 Behavioural and Micro-developmental Dimensions in the Water, Sanitation and Hygiene Sector During and After the Covid-19 Crisis

International organisations (such as the World Bank/UNICEF[1]) and other actors[2] in the field were quick to point out the link between Water, Sanitation and Hygiene (WASH) and COVID-19, as well as potential lasting effects. This subsection argues that the Covid-19 pandemic is likely to have behavioural effects, while further sharpening the policy and research focus on WASH. It also argues that this in turn might have downstream effects on other key development goals, such as improving child nutrition.

WASH infrastructure and behaviours are intricately linked to the management of the pandemic. Handwashing is one of the key measures against Covid-19, requiring both the possibility and habit do so. How was the pre-pandemic situation in developing countries? Figure 12.5 shows a simple time trend in the percentage of the population with basic handwashing facilities including soap and water, aggregated over low income countries in the period 2008–2017. In 2008, a mere 11.7% of people in low-income countries had such facilities. By 2017 this fraction almost doubled but remained at a low 19.6%. It will be interesting to follow and study these trends in a post-pandemic world; one can hope that Covid-19 has further underlined the need to increase this percentage both via the supply and demand side. On the one hand, policymakers and donors have now additional and very salient arguments for investment in the WASH sector. However, returns to investments in WASH in low-income countries have already been very high prior to the pandemic. For instance, UN Water (2021) cites a return of $5 for each $1 invested in the case of basic sanitation in rural areas. However, these returns have been primarily local. Given the global nature of Covid-19 and likely future pandemics, local investments in WASH will generate global externalities and benefits. On the other hand, individuals have been exposed to a public health information campaign of global proportions, highlighting the need of handwashing for instance. It remains to be seen to what extent all this has changed individuals' habits across geographies and subpopulations. One key dimension from a behavioural point of view are the establishment of more positive social norms, which are important in the WASH context (see for instance the work on India by Gauri et al., 2020). Given the global and complex nature of Covid-19 and information campaigns (and thus no clear counterfactuals),

---

[1] See for instance: World Bank, WASH (Water, Sanitation & Hygiene) and COVID-19, April 6, 2020, available at: https://www.worldbank.org/en/topic/water/brief/wash-water-sanitation-hygiene-and-covid-19 [Accessed 15 April 2021].

[2] See for instance: Stockholm International Water Institute, How is the WASH sector responding to COVID-19? Overview of COVID-19 Responses in 84 Countries, August 22, 2020, available at: https://www.siwi.org/latest/how-is-the-wash-sector-responding-to-covid-19-overview-of-covid-19-responses-in-84-countries/ [Accessed 15 April 2021].

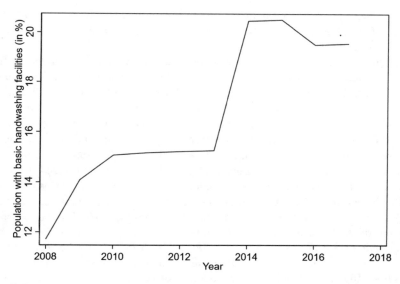

Fig. 12.5 Population with basic handwashing facilities (in %) in low-income countries (2008–2017)
Note: Own graph. Data are from the World Development Indicators curated by the World Bank and have been downloaded via *wbopendata* in STATA on March 17, 2021. The underlying source is WHO/UNICEF,Joint Monitoring Programme for Water Supply and Sanitation. The indicator code in *wbopendata* is SH.STA.HYGN.ZS ("People with basic handwashing facilities including soap and water (% of population)"). The aggregate for low-income countries is based on the World Bank's country classification.

as well as secular trends in hygiene practise, establishing causal effects will be challenging from a statistical point of view.

From a research and policy perspective, Covid-19 has further underlined the importance of behavioural economic and science insights. Key measures, such as regular hygiene, mask-wearing or vaccination campaigns, rely crucially on individual motivations, while having positive societal returns or externalities (for instance getting a vaccination shot can be modelled as pro-social behaviour, see Korn et al., 2020). Early on in the pandemic, numerous economists have highlighted the value of behavioural economics in fighting the pandemic. Behavioural economics advocates for nudges to address biases related to information avoidance, distorted risk preferences, present bias, social influence and the like (Soofi et al., 2020). The pandemic will therefore undoubtedly generate novel insights as to how to impact behaviour in the WASH sector and beyond. We argue that the pandemic will further magnify the trend in development economics as a discipline to depart from neoclassical models of decision making and to incorporate insights from psychology and related fields (for a review of behavioural development economics see Kremer et al., 2019). That said, effective behavioural interventions in WASH require comprehensive or systemic approaches. For instance, the Integrated Behavioural Model for Water, Sanitation, and Hygiene (IBM-WASH) by Dreibelbis et al. (2013)

rightly underlines that interventions need to take into account "*contextual*", "*psychosocial*" and "*technology factors*" at the levels of individuals, households, communities and society as a whole.

The prominence of role of hygiene in the pandemic period also matters for international development moving forward: WASH investments remain key to reaching the UN's Sustainable Development Goals (SDGs), in particular goals 2 and 6 6: "*End hunger, achieve food security and improved nutrition and promote sustainable agriculture*" and "*Ensure availability and sustainable management of water and sanitation for all*" respectively. Globally, about a fifth of all children under the age of 5 are still stunted, i.e. too small in stature for their age. Billions of people do not have the basic sanitary necessities needed for handwashing at home (see UN, 2021).[3] There is comprehensive and robust evidence that SDGs 2 and 6 are strongly associated with each other (for prominent examples in this broader literature see Spears, 2013, 2020; see also a review by Cumming & Cairncross, 2016). From a mechanistic point of view, Spears (2020, p.1) underlines that: "*…germs from feces cause diarrhea and other diseases, which can consume energy and harm the overall nutrition of growing children and of the mothers who nurture them in pregnancy and early life.*" To make this point at a very basic level, consider Fig. 12.6, where we plot the two bivariate associations between stunting rates of under-5 s in low and lower-middle-income countries with two key WASH indicators: the percentage of the population practising open defecation (not using a toilet) and having handwashing facilities, respectively. Note that indicators have been averaged over the period 2000–2017. Stunting is strongly and positively correlated with open defecation (corr = 0.366; p-value = 0.001). The slope (0.188, se = 0.049) of the linear fit suggests that moving from an open defecation rate of 23.6% (the sample mean) to zero, comes with a significant 4.4%-points decrease in stunting, which amounts to a decrease of 13.1% in stunting relative to its sample mean. Conversely, stunting is negatively correlated with handwashing (corr = −0.453; p-value = 0.000). The slope is −0.162 (se = 0.039) and implies that an increase in handwashing facilities from a sample mean of 35.5% to 100% is accompanied by a 10.5%-point decrease in the stunting rate, a sizeable 30.9% fall in stunting relative to its sample mean. Needless to say, these are correlations that may vary across geographies and contexts, and they also have not been adjusted for confounding factors and age-profiles as is done in the literature (Spears, 2013, 2020; see also Rieger & Trommlerová, 2016; Rieger

---

Fig 12.6 (continued) defecation and for stunting are WHO/UNICEF,Joint Monitoring Programme for Water Supply and Sanitation, as well as UNICEF/WHO,/World Bank, Joint child malnutrition estimates. The indicators codes in *wbopendata* are SH.STA.STNT.ZS ("Prevalence of stunting, height for age (% of children under 5)"), SH.STA.ODFC.ZS ("People practicing open defecation (% of population)") and SH.STA.HYGN.ZS ("People with basic handwashing facilities including soap and water (% of population)"). The sample is restricted to low-income and lower-middle income countries based on the World Bank's classification. Indicators have been averaged over the period 2000–2017, ignoring gaps in the time series. Sample size is 76 in the top figure and 65 in the bottom figure

---

[3] See goals and related statistics at UN (2021), as well as van der Hoeven and Vos (2021).

12 Covid-19 and Water

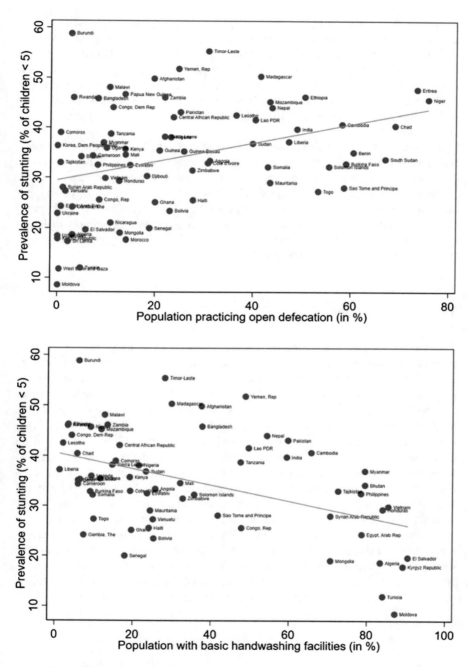

Fig. 12.6 Child Stunting, Open Defecation and Handwashing Facilities (Averages, 2000–2017) Note: Own graph. Data are from the World Development Indicators curated by the World Bank and have been downloaded via *wbopendata* in STATA on March 17, 2021. Underlying sources for open

et al., 2019). But these figures do drive home the basic point that these two SDG domains, WASH and nutrition, are empirically associated (Cumming & Cairncross, 2016). This in turn is relevant to the Covid-19 era. The UN (2021) advocates that Covid-19 is directly linked to both goals via food chains and hygiene. But the pandemic might also have downstream effects on the strong association between both goals in the short and medium term. If handwashing rates go up due to increased awareness and improved facilities, this may have positive implications for child health.

In sum, we hope that one positive side or aftereffect of the pandemic is that it may render the need for WASH investments and behaviour change more salient to individuals and policy makers around the world. Economic returns to WASH investments are global in times of a pandemic. Poor WASH conditions in developing countries will diminish global mitigation efforts of future pandemics. Local communities and the world as a whole need to be prepared for the inevitability of future pandemics. But WASH investments are equally vital to achieving the SDGs, in particular when it comes to child health (SDG 2).

12.6 Conclusions

The ongoing pandemic has brought renewed attention to issues of poor water governance. As handwashing and proper sanitation have become key tools in the fight against the pandemic, government officials and policymakers need to strengthen the provision of water services, especially in areas with chronic problems of water shortages. Properly-designed water policies can secure improvements in multiple fields: uninterrupted (safe) water provision, affordability of clean water for everyone, health benefits, food security and Covid-19 protection. This is, however, no easy task. It would require a rise in investment and international aid dedicated to water services, which may be unrealistic during a period of severe economic contraction and worsening public finances. Similarly, it would demand generous transfers of funds towards water utilities under financial strain and significant support towards vulnerable communities that lacked access to clean water much before the pandemic further reduced their incomes. The urgency is becoming, though, increasingly clear and some donors are already taking some steps in the right direction (as in the case of the Nordic Development Fund that provided $8.8 million to the African Water Facility in late 2020 for water supply and sanitation projects in arid parts of the Shale and the Horn of Africa to limit the spread of Covid-19.

References

Ahlers, R., & Merme, V. (2016). Financialization, water governance, and uneven development. *Wiley Interdisciplinary Reviews: Water, 3*(6), 766–774.

Akhmouch, A., & Kauffmann, C. (2013). Private-sector participation in water service provision: revealing governance gaps. *Water international, 38*(3), 340–352.

American Academy of Arts and Sciences (AAAS). (2019). *Water security meeting summary: Water in our future*. Notes of discussions at an American Academy Workshop, 19–20 June, 2019. The Battery Wharf Hotel, Boston, MA, United States of America. Unpublished.

Antwi, S. H., Getty, D., Linnane, S., & Rolston, A. (2020). COVID-19 water sector responses in Europe: A scoping review of preliminary governmental interventions. *Science of the Total Environment, 765*, 1–7.

Bayliss, K. (2014). The financialization of water. *Review of Radical Political Economics, 46*(3), 292–307.

Benton, T. G. (2020). COVID-19 and disruptions to food systems. *Agriculture and Human Values, 37*, 577–578.

Brinkman, H. S., de Pee, S., Aanogao, I., Subran, L., & Bloem, M. W. (2010). High food prices and the global financial crisis have reduced access to nutritious food and worsened nutritional status and health. *Journal of Nutrition, 140*(Suppl. 1), 153–161.

Butler, G., Pilotto, R. G., Hong, Y., & Mutambatsere, E. (2020). *The impact of COVID-19 on the water and sanitation sector*. International Finance Corporation, World Bank Group. Available at https://www.ifc.org/wps/wcm/connect/126b1a18-23d9-46f3-beb7047c20885bf6/The+Impact+of+COVID_Water%26Sanitation_final_web.Pdf. Accessed on 01 Apr 2021

Cheval, S., Mihai Adamescu, C., Georgiadis, T., Herrnegger, M., Piticar, A., & Legates, D. R. (2020). Observed and potential impacts of the COVID-19 pandemic on the environment. *International Journal of Environmental Research and Public Health, 17*, 1–25.

Chouchane, H., Krol, M. S., & Hoekstra, A. Y. (2018). Expected increase in staple crop imports in water-scarce countries in 2050. *Water Research X, 1*(10001), 1–7.

Cooper, R. (2020). *Water for the urban poor and Covid-19* (K4D Helpdesk Report 826). Institute of Development Studies (IDS).

Costinot, A., & Donaldson, D. (2012). Ricardo's theory of comparative advantage: Old idea, new evidence. *American Economic Review, 10*(3), 453–458.

Cotterill, S., Bunney, S., Lawson, E., Chisholm, A., Farmani, R., & Melville-Shreeve, P. (2020). COVID-19 and the water sector: Understanding impact, preparedness and resilience in the UK through a sector-wide survey. *Water and Environment Journal, 34*, 715–728.

Cumming, O., & Cairncross, S. (2016). Can water, sanitation and hygiene help eliminate stunting? Current evidence and policy implications. *Maternal and Child Nutrition, 12*(Suppl 1), 91–105.

Dreibelbis, R., Winch, P. J., Leontsini, E., Hulland, K. R., Ram, P. K., Unicomb, L., & Luby, S. P. (2013). The integrated behavioural model for water, sanitation, and hygiene: A systematic review of behavioural models and a framework for designing and evaluating behaviour change interventions in infrastructure-restricted settings. *BMC Public Health, 13*, 1015.

Falkendal, T., Otto, C., Schewe, J., Jägermeyr, J., Konar, M., Kummu, M., Watkins, B., & Puma, M. J. (2021). Grain export restrictions during COVID-19 risk food insecurity in many low- and middle-income countries. *Nature Food, 2*, 11–14.

FAO. (2018). *Water uses*. Food and Agriculture Organization of the United Nations. https://www.fao.org/nr/water/aquastat/water_use/index.stm#time. Accessed 19 Oct 2018

FAO. (2021). *FAOSTAT dataset*. Available at https://www.fao.org/faostat/en/#data. Accessed on 15 Feb 2021.

Gaddis, E., Grellier, J., Grobicki, A., Hay, R., Mirumachi, N., Mukhtarov, F., & Rast, W. (2019). *Freshwater policy. In global environment outlook-GEO-6: Healthy planet, healthy people*. UN Environment.

Gauri, V., Rahman, T., & Sen, I. (2020). Shifting social norms to reduce open defecation in rural India. *Behavioural Public Policy*, 1–25. In Press.

Hallam, D. (2018). Revisiting Prebisch-singer: What long-term trends in commodity prices tell us about the future of CDDCs. In *Background paper to the UNCTAD-FAO commodities and development report 2017, commodity markets, economic growth and development*. Food and Agricultural Organization of the United Nations.

Hanjra, M. A., & Qureshi, M. E. (2010). Global water crisis and future food security in an era of climate change. *Food Policy, 35*(5), 365–377.

IFC. (2020). *The impact of Covid-19 on the water and sanitation sector*. International Finance Corporation, World Bank Group. Available at: https://www.ifc.org/wps/wcm/connect/126b1a18-23d9-46f3-beb7-047c20885bf6/The+Impact+of+COVID_Water%26Sanitation_final_web.pdf?MOD=AJPERES&CVID=ncaG-hA. Accessed on 19 Feb 2021

Ingram, H. (2013). No universal remedies: Design for contexts. *Water International, 38*(1), 6–11.

Kalfagianni, A., & Papyrakis, E. (2021). Covid-19 and climate change. In E. Papyrakis (Ed.), *Covid-19 and international development*. Springer.

Klein, N. (2017). How power profits from disaster. The Guardian, 06 July 2017. Available at: https://www.theguardian.com/us-news/2017/jul/06/naomi-klein-howpower-profits-from-disaster. Accessed on 19 September 2021.

Kobayashi, Y., Heinrich, T., & Bryant, K. (2021). Public support for development aid during the COVID-19 pandemic. *World Development, 138*, 1–11.

Kolker, J. E., Erhardt, D., & Riley, S. (2020). *Considerations for financial facilities to support water utilities in the COVID-19 crisis*. World Bank Group. Available at https://elibrary.worldbank.org/doi/abs/10.1596/34043. Accessed on 01 Apr 2021

Korn, L., Böhm, R., Meier, N. W., & Betsch, C. (2020). Vaccination as a social contract. *Proceedings of the National Academy of Sciences of the United States of America, 117*(26), 14890–14899.

Kremer, M., Rao, G., & Schilbach, F. (2019). Behavioral development economics. In B. D. Bernheim, S. DellaVigna, & D. Laibson (Eds.), *Handbook of behavioral economics – Foundations and applications* (Vol. 2, pp. 346–393).

Laborde, D., Martin, W., Swinnen, J., & Vos, R. (2020). COVID-19 risks to global food security. *Science, 369*(6503), 500–502.

Malpass, D. (2021). *Covid crisis is fuelling food price rises for world's poorest. 29 January 2021, the Guardian*. Available at https://www.theguardian.com/business/2021/jan/29/covid-crisis-is-fuelling-food-price-rises-for-worlds-poorest. Accessed on 17 Feb 2021.

Mekkonen, M. M., Hoekstra, A. Y., & Becht, R. (2012). Mitigating the water footprint of export cut flowers from the Lake Naivasha Basin, Kenya. *Water Resources Management, 26*, 3725–3742.

Misra, A. K. (2014). Climate change and challenges of water and food security. *International Journal of Sustainable Built Environment, 3*(1), 153–165.

Mukhtarov, F. (2007). Integrated water resources management from a policy transfer perspective. In *International congress on River Basin management (March 22–24). Proceedings of the international congress on river basin management* (pp. 610–625). State Hydraulic Works. https://colinmayfield.com/public/PDF_files/Integrated%20Water%20Resources%20Management%20from%20a%20policy%20transfer%20perspective.pdf

Mukhtarov, F., Gasper, D., Alta, A., Gautam, N., Duhita, M. S., & Morales, D. H. (2021). From 'merchants and ministers' to 'neutral brokers'? Water diplomacy aspirations by the Netherlands – a discourse analysis of the 2011 commissioned advisory report. *International Journal of Water Resources Development*. https://doi.org/10.1080/07900627.2021.1929086

Neal, M. J. (2020). COVID-19 and water resources management: Reframing our priorities as a water sector. *Water International, 45*(5), 435–440.

OECD. (2020). *Green budgeting and tax policy tools to support a green recovery* (OECD policy responses to coronavirus (Covid-19)). Organisation for Economic Co-operation and Development. https://www.oecd.org/coronavirus/policy-responses/green-budgeting-and-tax-policy-tools-to-support-a-green-recovery-bd02ea23/. Accessed on 21 Feb 2021

Otero, G., Pechlaner, G., & Gürcan, E. C. (2013). The political economy of 'food security' and trade: Uneven and combined dependency. *Rural Sociology, 78*(3), 263–289.

Rieger, M., & Trommlerová, S. K. (2016). Age-specific correlates of child growth. *Demography, 53*(1), 241–267.

Rieger, M., Trommlerová, S. K., Ban, R., Jeffers, K., & Hutmacher, M. (2019). Temporal stability of child growth associations in demographic and health surveys in 25 countries. *SSM – Population Health, 7*, 100352.

Schmidt, J. J., & Matthews, N. (2018). From state to system: Financialization and the water-energy-food-climate nexus. *Geoforum, 91*, 151–159.

Soofi, M., Najafi, F., & Karami-Matin, B. (2020). Using insights from behavioral economics to mitigate the spread of COVID-19. *Applied Health Economics and Health Policy, 18*(3), 345–350.

Spears, D. (2013). *How much international variation in child height can sanitation explain?* (World Bank policy research working paper 6351). World Bank.

Spears, D. (2020). Exposure to open defecation can account for the Indian enigma of child height. *Journal of Development Economics, 146*, 102277.

Tortajada, C., & Biswas, A. K. (2020). COVID-19 heightens water problems around the world. *Water International, 45*(5), 441–442.

UN. (2021). Department of Economic and Social Affairs, *The 17 Goals*: https://sdgs.un.org/goals/goal2 and https://sdgs.un.org/goals/goal6. Accessed 17 Mar 2021.

UN Water. (2021). *Water, sanitation and hygiene*, Available at: https://www.unwater.org/water-facts/water-sanitation-and-hygiene/. Accessed 17 Mar 2021.

UNCTAD. (2020). *Impact of the COVID-19 pandemic on trade and development: Transitioning to a new normal*. United Nations Conference on Trade and Development.

van der Hoeven, R., & Vos, R. (2021). Reforming the international financial and fiscal system for better COVID-19 and post-pandemic crisis responsiveness. In E. Papyrakis (Ed.), *Covid-19 and international development*. Springer.

Watkins, K. (2006). *Human development report 2006-beyond scarcity: Power, poverty and the global water crisis*. UNDP Human Development Reports.

World Bank. (2020). *Supporting water utilities during Covid-19*. World Bank. Available at: https://www.worldbank.org/en/news/feature/2020/06/30/supporting-water-utilities-during-covid-19. Accessed on 19 Feb 2021

World Health Organization (WHO) and United Nations Children's Fund (UNICEF). (2017). *Progress on drinking water, sanitation and hygiene: 2017 update and SDG baselines*. World Health Organization (WHO) and the United Nations Children's Fund (UNICEF). Licence: CC BY-NC-SA 3.0 IGO. Available at: https://www.who.int/water_sanitation_health/publications/jmp-2017/en/. Accessed on 01 Apr 2021.

World Health Organization (WHO) and United Nations Children's Fund (UNICEF). (2020). *Global progress report on water, sanitation and hygiene in health care facilities: Fundamentals first*. Available at https://apps.who.int/iris/bitstream/handle/10665/337604/9789240017542-eng.pdf. Accessed on 01 Apr 2021.

Index

A
Absenteeism, 5, 120
Accountability, 152
Adaptations, 6, 8, 72, 122, 159
Adults, 51
Agricultural, 14, 34, 139, 152, 158–160, 163, 164
Aid, v, 6, 10, 40–42, 97, 152, 153, 165, 170
Amazon, 5, 137–144
Antibiotics, 104
Asia, 11, 12, 16, 23, 62
Austerity, 23, 99
Autocracy, 68
Automations, 4, 98, 99
Awareness, 17, 20, 31, 42, 56, 107, 153, 157, 170

B
Bacterial, 106
Bank lending, 33
Biodiversity, 141
Brexit, 30, 40, 69
Businesses, 11, 15, 30, 42, 50, 91, 95, 101, 104, 132, 150

C
Campaigns, 100, 105, 122, 124, 131, 132, 157, 166, 167
Capitalism, 98, 141–143
Carbon emissions, 6, 148, 149, 153
Cardiovascular, 106
Causality, 84

Children, 5, 48, 51, 53–57, 91, 92, 100, 104–107, 120–122, 125–127, 129–133, 166, 168–170
China, 1, 4, 11, 12, 30, 31, 35, 36, 39–41, 66–68, 107, 123, 142, 144, 148
Citizenship, 48, 67
Climate change, 2, 5, 6, 8, 147–154, 159, 161, 162, 164, 165
CO_2, 148–150
Colonial, 138
Commodities, 20, 97, 123, 140, 143, 163
Communications, 52, 54, 92, 120, 128, 152, 165
Commuting, 153, 165
Computers, 54, 128–130
Conditionality, 3, 10, 20, 21, 23–24, 42
Conflicts, 65, 69, 131, 141, 143
Connectivity, 129
Consumers, 154
Consumption, 14–16, 63, 64, 96, 153, 160, 165
Criminality, 122, 131
Crisis, v, 3, 4, 6, 9–25, 31, 33, 34, 36, 39–42, 47, 51, 56, 61, 65, 83, 90, 91, 98–101, 104, 106, 121, 125, 127, 133, 147–154, 158, 162, 164–170
Cultural, 33, 37, 46, 52, 56, 139, 141

D
Deaths, 1, 5, 31, 61, 62, 64, 68, 91, 103–106, 110, 113, 120, 123
Debts, 3, 10, 18, 19, 21–24, 42, 53, 100, 151
Decentralized, 113
Defecation, 168–169

© The Author(s), under exclusive license to Springer Nature Switzerland AG 2022
E. Papyrakis (ed.), *COVID-19 and International Development*,
https://doi.org/10.1007/978-3-030-82339-9

Deficits, 10, 23, 106, 123
Deforestation, 141, 149
Deglobalization, 30, 32, 33, 36, 42, 43
  economic dimension, 33
  political dimension, 33
  social dimension, 33
Deliveroo, 50
Democracies, 67, 69
Demographic, 87
Dental, 106
Dependencies, 6, 78, 158, 162–164
Depreciation, 19, 20, 22
Diabetes, 106
Dietary, 16
Digital gap, 5, 120, 128, 130, 153
Digital learning, 123, 127–130, 132
Digital technologies, 5, 7, 127, 128, 130, 132, 153
Distance learning, 124, 127, 128, 130, 132
Dreher, A., 32
Droughts, 47, 150

E
Earnings, 4, 19, 20, 50, 73, 76, 79–83, 85–87, 106
Ebola, 91, 105, 107, 111
Economic growth, 5, 64, 68, 123, 124, 164
Education, 2, 5, 7, 34, 50, 52, 54–56, 74, 90, 92, 99–101, 119–133, 151
Efficiency, 65, 149, 151, 152, 163
Egalitarian, 62
Electoral, 68, 69
Electricity, 96, 100, 150
Employment, 2–5, 7, 8, 11–14, 20, 33, 38, 46, 56, 60, 63, 64, 66, 68, 72–74, 76, 79–83, 85–87, 91–93, 98, 101, 119, 120, 125, 139, 149, 165
Environmentalists, 6
Equalities, 42
Equilibrium, 15, 91
Equity, 65, 108, 110–112
Ethnic, 5, 8, 148
Exports, 19, 20, 33–37, 63, 143, 162–164
Externalities, 15, 149, 166, 167
Extractive industries, 5, 138, 140–141
Extractivism, 143

F
Facemasks, 46
Farming, 95, 164
Females, 49, 53–55, 125

Fertility, 105
Financial assets, 21
Financialization, 158, 160
Financial support, 3, 5, 18, 20, 23, 40, 41, 100, 113
Financial system, 2, 42
Fiscal stimulus, 15, 18, 23, 150, 151
Fiscal system, 3, 9–25, 65
Floods, 150
Foods, 2, 6, 10, 15–16, 24, 50, 51, 96–98, 100, 130, 140, 158, 162–164, 168, 170
Foreign Direct Investment (FDI), 33, 34, 36–37, 41, 42, 77

G
Gender, 4, 48, 53, 56, 74
Geopolitical, 40, 142, 144
Globalization, 29–43
Governance, 4, 6, 42, 60, 68–69, 131, 157–162, 170
Governments, v, 2, 3, 6, 11, 17, 18, 20, 21, 23, 38, 39, 50, 56, 57, 63–65, 71, 72, 92, 96, 97, 100, 101, 104, 121, 123–127, 130–133, 141, 143, 149, 151, 154, 159, 160, 164, 165, 170
Gygli, S., 32

H
Handwashing, 46, 90, 94, 165–170
Harmonisation, 121
Health, v, 1–8, 10, 11, 17, 18, 20, 24, 25, 30, 31, 34, 40, 41, 46, 48–54, 56, 59, 60, 64, 68, 71, 72, 90–92, 98–100, 103–113, 119, 120, 123, 124, 127, 130, 132, 137, 138, 140, 141, 144, 150, 152, 158–161, 165, 166, 170
Healthcare, 20, 152, 158, 159
Heteroskedasticity, 78
HIV, 5, 105, 106, 113
Hospitals, v, 5, 48, 49, 54, 104, 106–112, 114
Households, 4, 5, 7, 14, 15, 17, 51, 53–56, 63–68, 91, 95–97, 100, 120, 124, 125, 128–131, 133, 148, 157, 159, 168
Hygiene, 90, 94, 157–159, 166–170

I
Illiteracy, 154
Immunization, 41
Imports, 6, 20, 33–35, 63, 162–164
Income losses, 2, 11, 12, 14, 15, 165

Indebtedness, 19
Indigenous communities, 5, 6, 8, 138–142, 150
Inequalities, v, 2, 4, 7, 10, 14, 16–21, 24, 42, 51, 59–69, 92, 100, 120, 121, 124, 128–130, 148, 150, 152
Influenza, 62, 91, 104
Informal economy, 2, 4, 7, 89–97, 100, 101, 150
Informality, 86, 90–92, 94, 98, 100, 101
Information, 1, 2, 4, 7, 8, 15, 30, 33, 46, 50, 53, 54, 56, 64, 77, 92, 108, 111, 112, 120, 122, 123, 127, 128, 130, 131, 147, 150, 166, 167
Infrastructures, 1, 2, 91, 107, 111, 127–129, 132, 133, 139, 147, 150–152, 158, 159, 161, 164, 166
Institutions, 21, 33, 39, 41, 50, 51, 56, 68, 69, 139, 159, 164
Integration, 30, 33, 34
Interconnectedness, 30
Intergenerational, 138, 141
International Monetary Fund (IMF), 3, 11, 12, 15, 19, 20, 23, 24, 42, 124, 149
Internet, 37, 43, 50, 74, 98, 100, 101, 127, 128
Interventions, 6, 46, 57, 65, 97, 99–101, 107, 113, 123, 126, 154, 158, 167, 168
Interviews, 3, 46
Isolationism, 2

L
Labour, 4, 7, 11–15, 30, 38, 39, 60–64, 68, 71–87, 93, 125, 153, 165
Latin America, 5, 11, 12, 23, 67, 119–133, 137, 142, 143
Lending, 20, 21, 23–24
Liquidity, v, 3, 10, 20, 21, 24
Livelihoods, 3, 10, 17, 20, 23, 30, 48, 49, 90, 95–97, 104, 130, 141, 151
Lockdowns, 5, 7, 11, 18, 31, 38, 48, 50, 51, 53, 54, 56, 63–65, 72, 86, 90, 92, 104, 105, 120, 123, 124, 127, 131, 133, 140, 148, 165

M
Macroeconomics, 17, 23, 60, 63–66, 91, 158, 164–165
Malaria, 5, 105, 106, 113
Malnutrition, 106, 162, 168–169
Manufacturing, 34, 99, 164
Marginalization, 138–141

Marriage, 105
Maternal, 104, 106
Media, v, 10, 56, 79, 83, 86, 120, 124, 125, 132, 133, 152, 170
Medication, 105
Medicines, 6, 106, 140, 141
Mental health, 49, 106, 107, 153
Migrants, 3, 4, 38, 39, 45–57
Migration, 2, 33, 37, 39, 47, 99, 150
Minorities, 5, 8, 69, 100
Mitigation, 6, 8, 15, 133, 151, 170
Mobility, 2, 3, 6, 7, 71, 124, 148, 154
Modernisation, 123
Mortality, 2, 62, 64, 68, 72, 105, 106, 111, 138
Mothers, 5, 48, 53, 55, 104, 168
Multilateralism, 10, 25
Mutations, v, 2, 5, 103, 113

N
Neonatal, 104, 105
Neurosurgical, 106
Nutrition, 24, 106, 130, 166, 168, 170

O
Obesity, 106, 148
Ocean acidification, 150
Oligopolistic, 21
Ophthalmic, 106

P
Pandemics, v–vii, 1–10, 14–17, 19–25, 30–36, 38–42, 46–52, 55, 56, 59–69, 71–74, 76, 82, 83, 85–87, 90–92, 94, 95, 97–101, 104–107, 110, 111, 113, 119–121, 123–125, 127, 128, 130–133, 137–139, 141–143, 147–154, 157–161, 164–168, 170
Parenting, 55
Parents, 54–56
Patents, 17, 37, 142
Patients, 5, 105–108, 110, 113, 123, 131
Pharmaceuticals, 2, 17, 100, 142, 143
Plutocracy, 69
Policymakers, 99–101, 166, 170
Politics, 151
Populists, 69
Poverty, v, 2, 5, 7, 10, 14–16, 24, 25, 42, 53, 60, 65, 89, 111, 122, 124, 125, 130, 132, 138, 150, 154
Precariousness, 4, 7

Pregnancies, 105, 168
Private sectors, 22, 33, 154, 159
Profits, 10, 21, 90, 91, 142
Protectionism, 25, 30
Psychological, 49, 107, 133, 154
Public sector, 20, 121

R
Race, 142, 150
Racism, 138, 139
Radios, 127
Recessions, 4, 10–12, 14–16, 18, 19, 23, 30–32, 36, 59, 64, 65, 86, 98, 162
Reforms, 3, 10, 21–25, 42
Refugees, 46, 49, 53, 54, 56, 148, 157, 159
Religious, 52, 56, 113, 121, 154
Remittances, 11, 15, 33, 37–39, 41, 42
Remoteness, 5, 144
Renewable, 165
Resilience, 5, 10, 31, 32, 42, 138–144, 152
Restaurants, 50, 51
Robots, 99
Rural, 16, 51, 122, 128, 130, 151, 163, 166

S
Sanctions, 77
Sanitary, 90, 132, 168
Sanitation, 6, 49, 94, 157–160, 165–170
Sanitizers, 36, 46
Savings, 14, 50, 60, 63, 64, 83
Scarcities, 4, 6, 91, 142, 157, 162, 163
Schooling, 7, 53, 54, 56, 120–123, 125–132
Securitisation, 159–161
Services, 2, 4–7, 14, 19–22, 33, 34, 36, 46, 47, 50, 51, 56, 63, 64, 98, 100, 105, 111, 113, 121, 130, 137–139, 141, 148, 153, 154, 158–161, 170
Shopping, 98, 153
Sickness, 4, 72, 123, 148
Simulations, 65, 149
Socialization, 130
Solidarity, v, 7, 20, 48, 56, 130
Stigmatisation, 107
Stunting, 168–169

Sub-Saharan Africa, 11, 12, 16, 23, 41, 159
Sustainability, 152
Sustainable Development Goals (SDGs), 3, 7, 10, 21, 24, 25, 97, 168, 170

T
Tariffs, 33, 159, 163
Tax, 3, 10, 21–22, 42, 63, 150, 165
Teachers, 55, 119, 131–133
Telemarketing, 99
Teleworking, 14, 153
Tourism, 3, 7, 33, 34, 38, 42, 165
Trades, 2, 3, 6, 7, 11, 14, 15, 18, 30–36, 41, 42, 61, 63, 69, 104, 142, 163, 164
Training, 96, 98, 101, 128, 151
Transfers, 65, 96, 97, 99, 101, 113, 160, 170
Tuberculosis, 5, 105, 106, 113

U
Unemployment, 5, 7, 12, 13, 46, 50, 63, 76, 79, 81–83, 85–87, 89–91, 96
Unsustainable, 6, 148, 153
Urbanisation, 149, 158, 164

V
Vaccines, v, 2, 3, 17–18, 30, 31, 40, 41, 59, 100, 103, 113, 123, 141–144, 154
Ventilators, 36
Violence, 53, 61–62, 131
Vulnerabilities, v, 2, 4, 7, 42, 48, 51–53, 55, 57, 138–141, 143, 147, 164

W
War, 61, 62
Water, 5–8, 49, 90, 94, 100, 139, 148, 157–170
Wealth, 21, 22, 60–63, 67, 68, 143
Women, 4, 5, 7, 48, 53–54, 56, 90, 91
Workers, 4, 5, 11, 12, 14, 15, 24, 34, 51, 63, 64, 72, 86, 89–92, 94–101, 107, 113, 130, 139, 140
World Bank, 3, 15, 16, 20, 41, 66, 94, 95, 125, 159, 164, 166–169

Printed in the United States
by Baker & Taylor Publisher Services